W9-CNA-568

creative ideas to organize your home

FREE GIFT!
EASY-STITCH

creative ideas to organize your home

50 step-by-step projects to bring order into your life

LINDA PETERSON

CICO BOOKS
LONDON NEW YORK

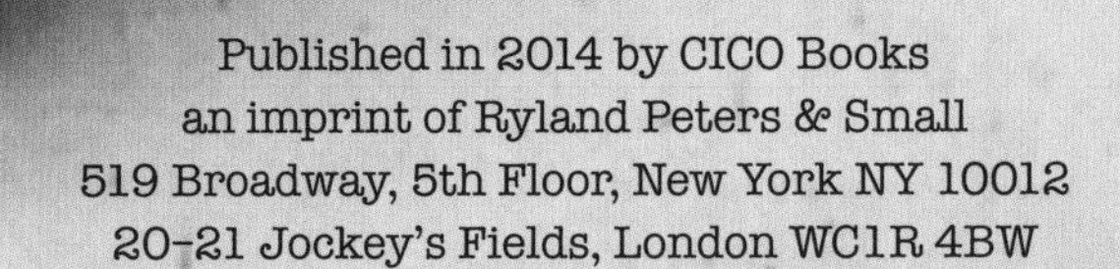

Published in 2014 by CICO Books
an imprint of Ryland Peters & Small
519 Broadway, 5th Floor, New York NY 10012
20–21 Jockey's Fields, London WC1R 4BW

www.rylandpeters.com

10 9 8 7 6 5 4 3 2 1

A CIP catalog record for this book is available from the Library of Congress and the British Library.

ISBN: 978 1 78249 097 5

Printed in China

Editor: Marie Clayton
Design: Mark Latter
Project photography: Gavin Kingcome
Stylist: Luis Peral Aranda
Step-by-step photography: Geoff Dann

contents

introduction

Dear Clutter Keeper friend of mine, and yes, I'm talking to you—you have stuff laying around everywhere: piles, stacks, books, pencils, paper, art supplies, bathroom toiletries, this, that, and the other... am I right? Well don't think you're the only one! Trust me, I fit the above description just as much as anyone until I took control!

My Mom used to say to me, "Linda, a place for everything and everything in its place!" That works great if you have a place for everything and not so great if you don't. Unfortunately, because I collect stuff, I'm more familiar with the latter. So, that leaves me two options: get rid of the stuff—either throw it away or give it to charity—or let stuff sit out and about and become overwhelming clutter. I'm not talking about stuff that is pure trash. I'm talking about useful stuff, things that we need in our daily life, or things that we need on a very frequent basis.

If you are a creative person by nature, you are probably horrified by the first option of throwing away and so the first option is NOT an option, most of the time. That leaves the latter, a life of stuff piling up, because you can't bear to throw anything away and you reason that someday, somehow you will need it—and when you need it, you will have it and that is why it's worth

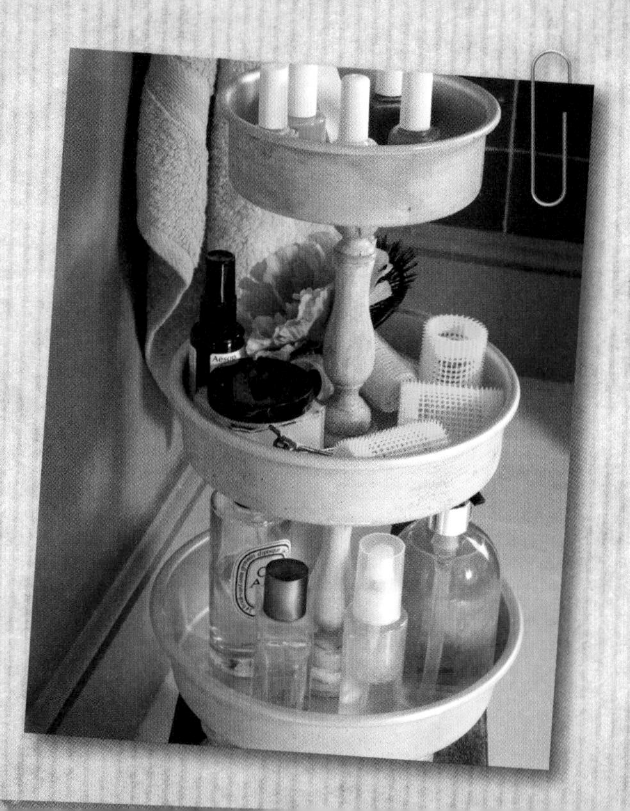

keeping! Sound like you? I know my husband will read this and know that I have completely described myself. It's frustrating for right-brained people to be neat and exact and meticulously tidy, though not impossible. Years ago when my daughter was in elementary school, I commented to one of her teachers—"Wow, you know where everything is at, you are so organized, I wish I could be like that." And her response? "Well, you can't be—you're creative! It's just not in your nature." Well, part of that is right, creativity is in my nature, but I CAN be organized—I just I have to find ways that I can do it creatively.

That is the inspiration and theme of this book. I'm not here to help you organize your entire life, your closet, your kitchen cupboards—no, I'm here to help you organize the little cluttery things that pile up, and to find creative solutions to keep them neat and tidy. We know that books go on shelves. We know that we can stuff things in baskets and forget about them until later. We clean out the bins and they stay that way for a while but, sooner or later, we are on the clutter merry-go-round again.

The projects I've selected and created for this book all had to meet certain criteria; they had to be artful or creative, use easily sourced materials (preferably ones that were free), and they had to be things that I would use on a permanent basis, not something I would use for a while and then kick to the side—practical solutions that really worked for the long haul. As you peruse the projects in this book, you'll soon realize that your local hardware store is really a craft haven. You'll learn to use materials in new and repurposed ways and you'll have fun in the process while de-cluttering your workspaces.

You may be frustrated with your lack of space, especially if you live in a big city. But the reality is that you have space. There is space on countertops, under the bed, beside the bed, and how about this one—air space? What do buildings do in big cities with little space: GO UP! So, think vertical—as in the Three Tier Fountain Storage (see page 48), take advantage of air space. Have a look around and see hidden areas of space you can discover and then take full advantage of them.

Many of the projects have multiple uses—for example the Art Supply Center on page 80 could easily double as a drink caddy. You could alter the idea and color scheme to fit a backyard party. Or how about the Folding Travel Journal on page 130—it could double as a receipt holder or maybe even a coupon organizer, or perhaps it's simply a fold up card that you'd like to give to someone special with all sorts of sweetness tucked inside.

I'd like to encourage you, too, to think beyond the projects and the color schemes. I've tried to show a variety of styles but mostly I've used my personal style, which is vintage, upcycled, and grunge. If you are more contemporary, you can easily simplify and adapt these projects to fit your décor by simply changing the color to suit your taste, not antiquing, and keeping embellishments minimal—this will give the finished item a simple, more contemporary feel.

I also don't want you to think you have to rush out and buy loads of stuff to organize. This book is designed to be budget friendly and the ideas are easy to make—even if you are a beginner. However, I would invest in a good quality battery-operated drill (or borrow one from your significant other when they're not looking). You will be glad you did and having this quality tool will make the whole organizing experience much more enjoyable.

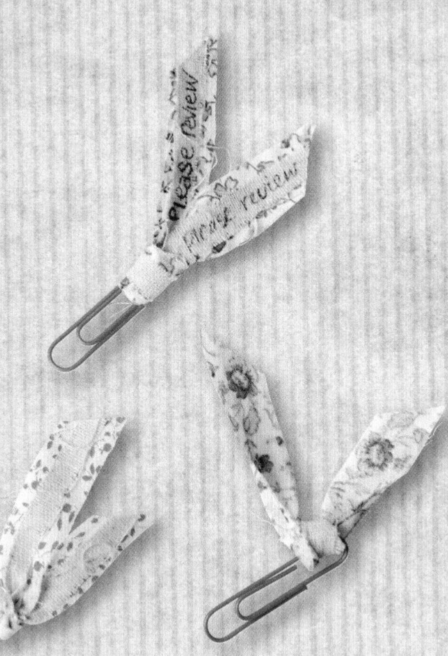

I personally believe that if we invest a little time and effort into these organizational projects we will be more inclined to keep things that way, be less stressed and happier, and—in the long run—will free up time for activities we really want to enjoy and do.

So, here's to our organizing journey—we're in this together my friend...

Linda

PARIS
COMPAGNIE D'ASSURAN

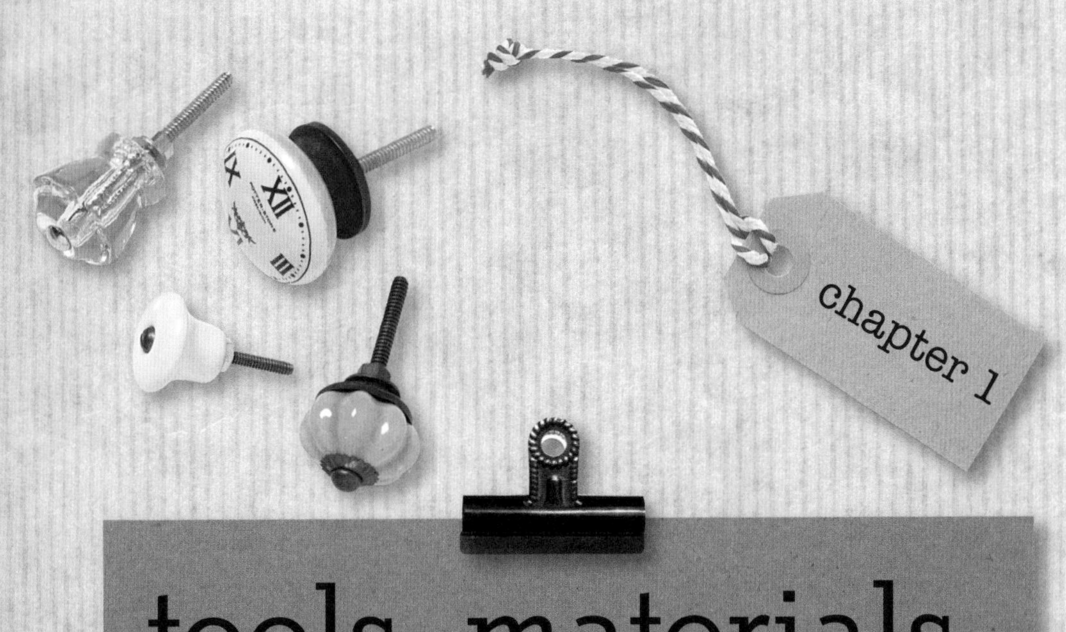

tools, materials, and techniques

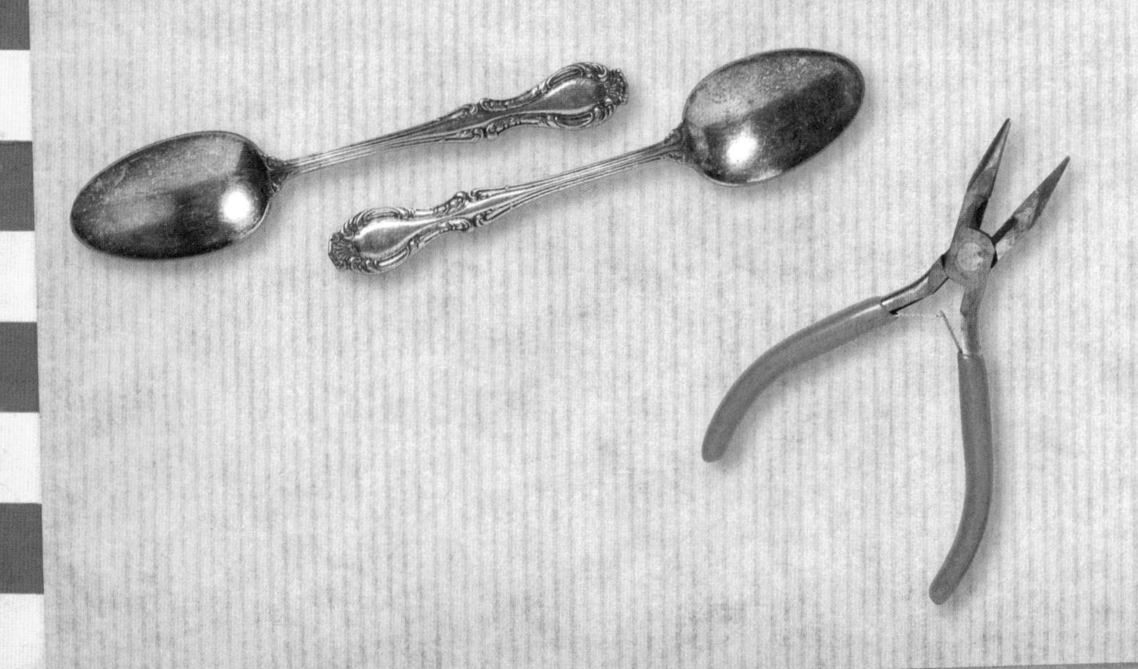

tools & materials

Nothing is fun if it's not easy! I have tried my best to use tools and materials that are easily sourced; most likely you have many of these items already. As with any project you undertake, the right tools for the job make your work much more enjoyable. I invested in a very nice battery-operated drill with quality drill bits, which has been worth its weight in gold!

General tools

Basic pliers All-purpose pliers to be used with metal and wire. The jaws may have teeth that allow you to grip and pull wire tight, so be careful that these do not leave marks on the softer materials.

Standard wire cutters To cut wire to length, or remove excess wire.

Drill and bits An electric drill will make drilling holes faster and easier, but you can also use a hand drill. Drill bits come in a range of sizes—I mostly use ¼ in. (6 mm) in these projects.

Vise A vise will hold small pieces secure for cutting or drilling—the type shown here has a head that can be adjusted to hold items at different angles.

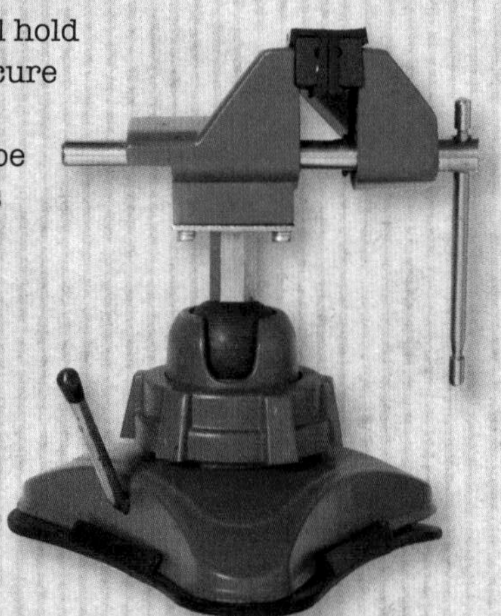

Center punch Used to create an indentation in metal before beginning to drill a hole so that the bit does not slip out of place.

Heat gun This gets hotter than a normal hairdryer. I use it to speed up the drying process in many of my projects.

Scissors Heavy-duty scissors (red handle) can cut thin pieces of wood and metal.

General all-purpose scissors (orange handle) are used to cut paper and thin cardstock.

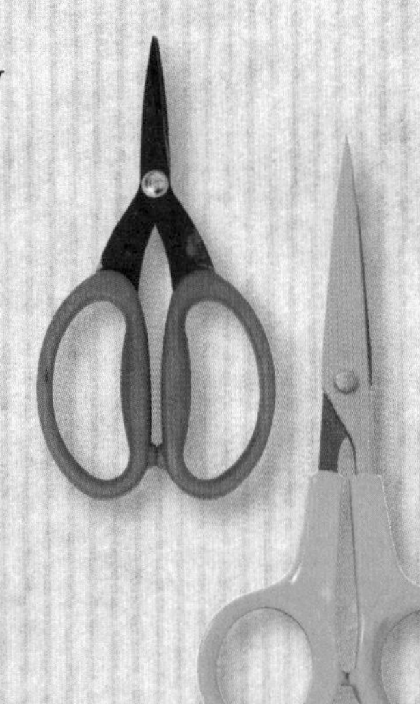

Cutting tools

Crop-A-Dile A useful scrapbooking tool that is designed to punch holes into a variety of materials. It will also set small eyelets.

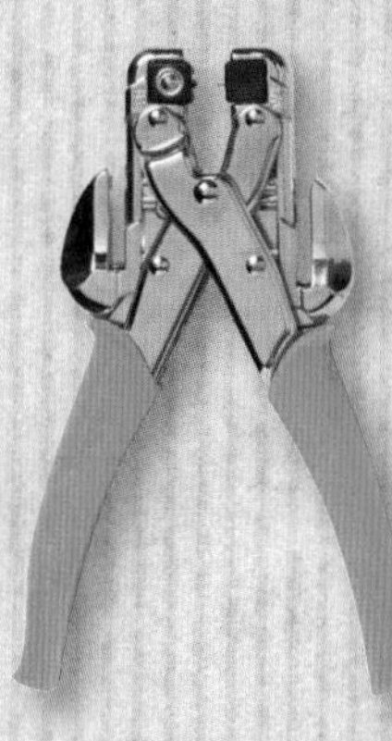

Fabric scissors Specialty scissors designed for fabric. You must never ever cut paper or any other material with these scissors or you will dull the blades quickly.

Needle tool A tool with a sharp point at the end. I use it instead of a punch when I need to make a small hole.

Hand drill and small drill bits A very handy miniature drill that will hold very tiny drill bits.

Craft knife A very sharp knife is useful in cutting a variety of materials, from thick cardstock to foam core. It is very successful in cutting your fingers too—so be careful and keep them tucked out of the way (yes, I've learned the hard way!).

Leatherman multi-tool This one tool does it all. It's not the most durable, but it has nearly every kind of tool you will need right at your fingertips. With pliers to pocketknife, and even small screwdrivers, it is really handy to have around.

Miter box To use with a hacksaw to cut wood. Diagonal cuts can also be made with this box ensuring that all your cuts are at exactly the same angle and size.

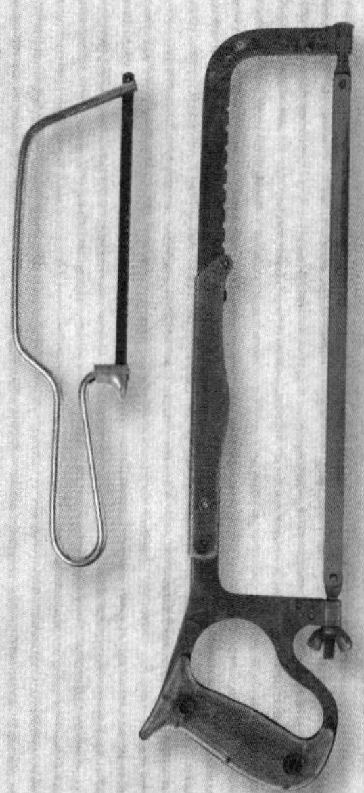

Hacksaws I used this type of saw for all the projects requiring sawing in the book. They are easy and safe to use and don't require electricity—just a little elbow grease!

Spellbinders® Artisan X-plorer™ die cutting machine and dies This is a mini die cutting machine used with metal dies to cut a variety of shapes. It will cut various thicknesses of paper, fabric, and metal into intricate shapes by passing them through the machine.

Sanding and coloring tools

Metal file One of my staple tools. I use it to file burrs off metal. These come in a variety of sizes and shapes to fit even into small areas.

Sandpaper sheets Be sure to have several grits available. The smaller the number the more coarse the grit is. I like to have a selection of 120 to 400 when working with woods and metals.

Sandpaper block This is sandpaper in block form. You can purchase these with different grits on each side. They are very convenient and useful, especially if you just need to touch up a small area. I use these a lot to distress and age pieces to give them a vintage look.

Color wash Available at craft stores and generally found in spray bottles. This thin solution is water soluble and translucent. It can be used on wood and fabric. You can even make your own (see page 29).

Acrylic paints Water-based paints that come in a variety of colors and consistencies. Often DIY stores will sell samples of their latex paint in sample sizes—this is a great way to get your own custom colors blended in small amounts.

Brushes I use general grade brushes—nothing fancy since much of the time I am applying faux finishes to things and aging them. The most important thing to me is to have a variety of sizes and to keep them clean.

Toothpicks/cocktail sticks I find these tiny sticks have so many uses, from applying small amounts of glue, to using as fillers to get me out of a jam when I've drilled a hole a little too big.

Ink dye For aging and distressing. I use a chocolate color, but ink comes in many colors so don't be afraid to experiment to match your décor.

Cosmetic sponges These are so handy and double as brushes to me. I use them to apply ink, glues, and to paint with. When the ends of the sponges get dirty or ragged, I trim them off with scissors and get a lot of mileage out of them.

Adhesives and tapes

Tacky dot tape Little tacky dots of glue that are useful to hold things together temporarily.

Fabric glue Useful to stick fabrics together. I use this as a substitute if I don't want to get out my sewing machine and stitch something together. There are different types of fabric glues; some are washable and some are not, so make sure you read the label.

Basting adhesive A type of fabric glue that has a quick tack time. It will hold pieces of material together without using pins so that you can sew the pieces together. This type of glue will generally wash away.

A Helpful Hint: There are so many different kinds of adhesives. Be sure that you are using the appropriate adhesive for the material you are using. This is so important if you want to achieve a strong bond.

Glass and bead glue It is important to use this if you are sticking anything to glass because only certain glues will maintain a bond.

Two-part epoxy glue This type of glue comes as a resin and a hardener; when the two are mixed together the reaction forms a very strong permanent bond. It comes in a variety of cure times—a 5-minute cure time is the one that I use the most. Be aware that this type of glue can yellow over time.

Fast grab tacky glue A general all-purpose thick white glue that sticks things together quicker than regular tacky glue.

Wood glue Use to create a permanent bond on wood. You can use it in combination with nails or screws. It is difficult to clean up once it begins to cure so make sure you wipe away excess quickly.

All-purpose glue A thick white glue used for general sticking. It's a staple in my studio.

Decoupage medium A clear thin glue that I use to stick paper to surfaces. When used as a top coat it acts as a good sealer and makes the material somewhat water resistant.

Double stick tape Tape that is tacky on both sides—it's a "can't live without" kind of tape for me. It comes in a variety of widths, which make it handy for all sorts of projects.

Double stick sheet As double stick tape but in sheet form. It comes in a variety of sizes.

Magnetic sheet Magnets that come in sheets. Some of them have an adhesive back already applied.

Magnetic dots As magnetic sheets but in dot form. Magnets come in many shapes and sizes and also have varying degrees of grip.

Hook-and-loop dots A two-part tape with one side consisting of soft loops and the other of somewhat scratchy hooks. When the two are put together they grip to one another but can be pulled apart when required again and again.

Patterned Duck Tape® The same type of tape as standard Duck Tape but in a variety of printed patterns.

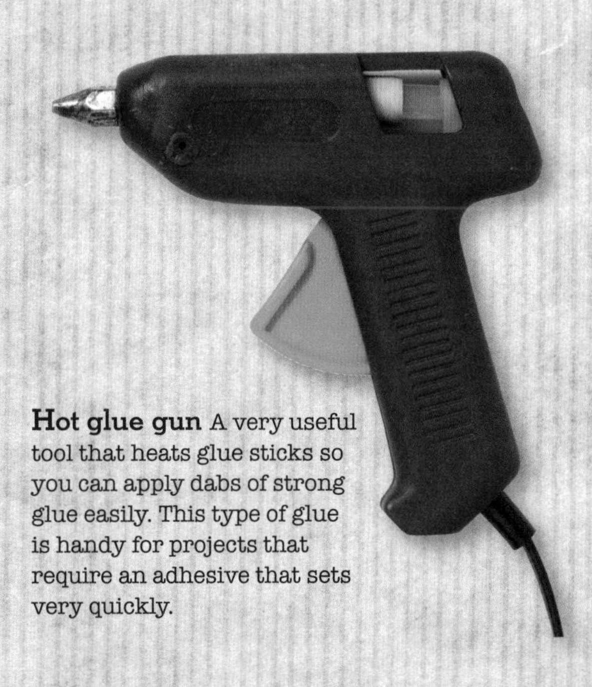

Hot glue gun A very useful tool that heats glue sticks so you can apply dabs of strong glue easily. This type of glue is handy for projects that require an adhesive that sets very quickly.

Filler materials

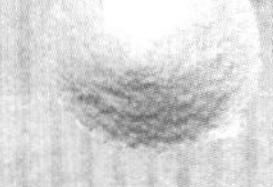

White sand Available in a variety of grits and colors. Usually found in the floral departments of craft stores.

Crushed shells Generally used to fill glass bowls for flower arrangements, but we use it as a decorative element and a way to add weight for more stability in some projects.

Dried beans Easy to acquire and inexpensive. Beans come in different colors and sizes and add a nice texture as a filler.

Casting materials

Silicone molds Often found near soap-making or candle-making supplies, silicone molds come in lots of shapes and sizes. Even though silicone molds may be a little more expensive than others, they are nice because objects release easily from them and they clean up readily.

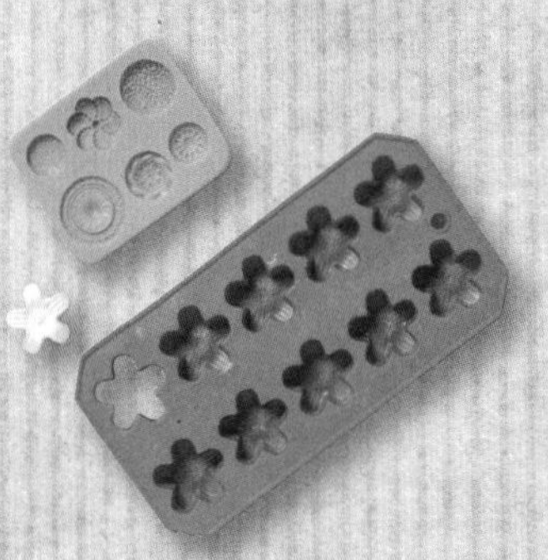

Silicone ice cube flower tray Many ice cube trays are made of silicone and offer fun shapes. Don't forget to check out your kitchen supply store and see what they have to offer.

Cool2cast™ fiber plaster casting medium A casting medium made of plaster and fiber. When mixed with water, it turns into a thick liquid similar to pancake batter. It can be poured into molds and will harden on its own.

Wooden base materials

Popsicle sticks Easily found in craft stores, I use these as decorative elements and as tools when applying glue.

Clothes pins Not just for hanging clothes out to dry—these little clips can be decorated and used in lots of ways to keep your papers and notes under control. They also make great mini-clamps to keep pieces together while glue is drying.

Dowel rods Available in craft and hobby stores in a variety of circumferences and lengths. They can be painted and stained. You can even find knobs to fit the ends to finish them off.

Wooden rules Often found in the school supply section of your local stores, wooden rules add a fun and decorative touch to many of the projects in the book. I use these because they are so inexpensive and easy to find.

Paint stirring sticks If you ask at your local DIY store they will probably give you a handful for free. Generally made of wood, they can be cut to size and decoupaged. Thinner ones can be cut with heavy-duty scissors.

Wooden candlesticks I use these instead of spindles. Readily found in the woodcraft section of your local craft store, they can also be cut to size with a hacksaw.

Wooden yardstick Not just a great measuring device, but useful to cut down to create small wooden projects. I also like the "ruled" pattern on the wood, which gives it a wonderful accent and texture.

A Helpful Hint: Also useful are 2-in. (5-cm) wide sections of wood—I save all of my scrap lumber, from blocks to different odd lengths. Small blocks of wood make great bases for knobs. If you don't have scraps of wood already, DIY hardware stores sell small pieces of wood and many will cut larger pieces of wood to size for you at very little or no additional charge.

Texture and pattern materials

Decorative pattern papers Scrapbooking papers come in a variety of patterns and colors and decoupaging them to wood is a great way to cover the surface. You can also tear sheets from old books and magazines for decoupage.

Chicken wire Once you work with chicken wire, you'll love it because it is so moldable and pliable to create nearly any shape. Chicken wire can also be rusted or painted and cuts easily with wire snips. A word of caution: be careful because the cut ends can be rather sharp and may scrape the skin.

Twine A thick rope that is strong and durable. I love to decorate and embellish with twine because it lends itself to vintage and rustic looks.

Fabric strips These add a soft touch to projects or give a shabby chic feel, and look great when combined with vintage finds.

Rickrack braid/trim A zigzag trim available in a variety of widths and colors to add a touch of whimsy to your projects.

Buttons These come in a variety of patterns, shapes, sizes, and finishes and are an easy way to embellish your projects. I love to use buttons as a center for whimsical flowers.

Beads Girls and beads go together ...right? There are so many shapes and colors and textures of beads—you can't go wrong adding beads and bling to your projects.

Thread General-purpose thread used in sewing projects. For a decorative touch, select a contrasting color to your fabric.

Fabrics I always have on hand a variety of fabrics, including calico, homespun, organza, and tulle. I have a range of textures, patterns, and prints. Since I don't sew a lot of large items, I buy fat quarters—this gives me a lot of color options without taking up too much space.

Raffia A grass-like fiber that is generally found with floral supplies. I use it as a decorative element—it comes in a variety of colors and you can dye it with color wash.

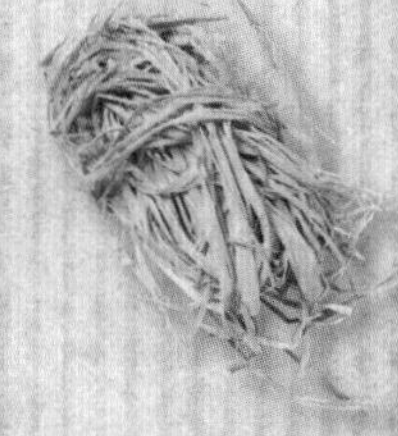

Ribbon I always a staple supply in my studio. Keep a variety of ribbons with different colors, patterns and widths handy at all times.

Straight pins Used for pinning fabrics together.

Other materials

Corks Save corks from wine bottles or pick up a few inexpensively from a winemaking supply store. Corks are useful for all sorts of things: making bulletin board pins, carving shapes for stamping—and in this book, I use them to finish off the ends of PVC pipe to give it a nice decorative touch.

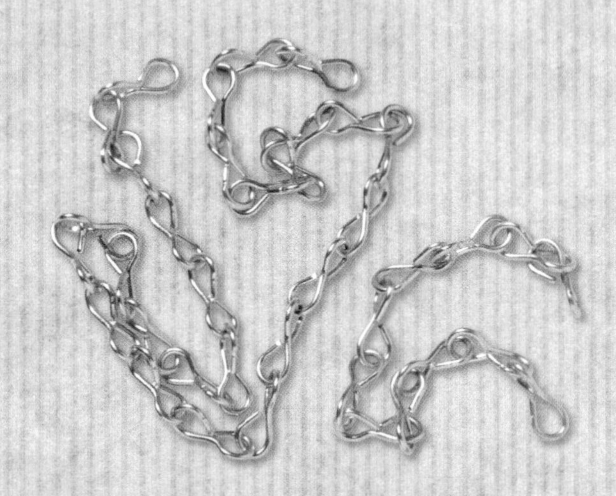

A Helpful Hint: *As you can see from these pages, many things that you would normally throw in the trash are actually useful materials that can be used in all kinds of ways. Before you throw anything away, stop and think if you could recycle or upcycle it instead!*

Jack chain This is a type of chain with little figure eight shaped links that I've used often in this book, and for a lot of my DIY home projects. It is sold in a package or by the length—DIY stores make it easy to purchase just the length you need. I like to keep several sizes and patterns on hand. Chains can also be hung from the ceiling with hooks allowing you to make use of air space.

Bottle caps A great thing to recycle! Bottle caps can be upcycled into magnets, they are easy to paint any color and can also be purchased reasonably cheaply from wine and beer making supply stores.

Glass pebbles or baubles These flat back glass pieces can typically be found in the floral departments of craft stores. They come in a variety of shapes and sizes as well as colors. I always keep clear ones on hand.

Fender washers These differ from regular washers because you'll notice that the outside rim is fairly thick compared to the smaller hole in the center. They help to distribute weight and provide a secure anchor. They can also be used as spacers. When using fender washers, make sure that the dowel screws or nuts and bolts you use with them are the correct size.

Picture-hanging wire Attach picture-hanging wire between small brass screw eyes to hang pictures on the wall.

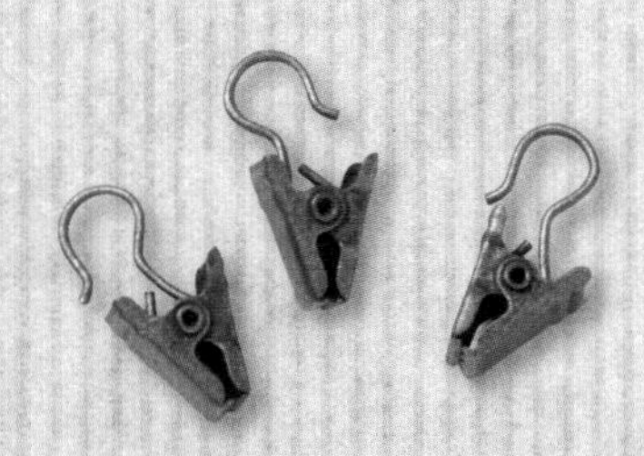

Miniature hanging Hercules clips Super cute little clips that make it easy to hang things. I've used these to hang fabrics, pictures, and more. They come in a variety of sizes and can be found at office supply and DIY stores.

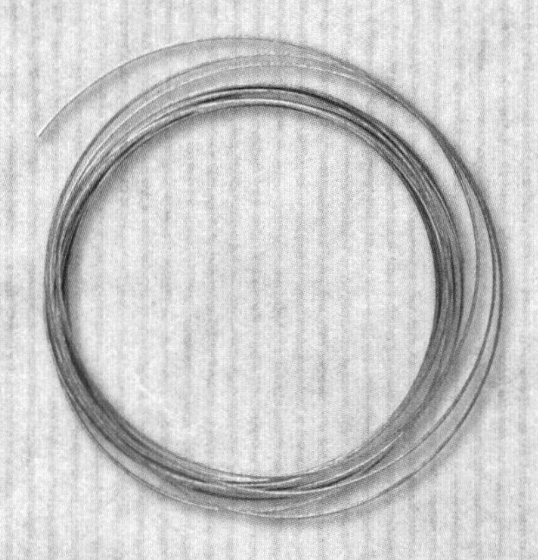

Picture wire hooks
These lightweight hooks are great to hang things that are not too heavy, such as picture frames. If you need something more durable, I suggest screw eyes (see page 20).

Wire The heavy-duty wire I use comes in spools and is relatively inexpensive. It can be cut easily with durable wire snips and has all sorts of uses such as creating hooks or attaching things.

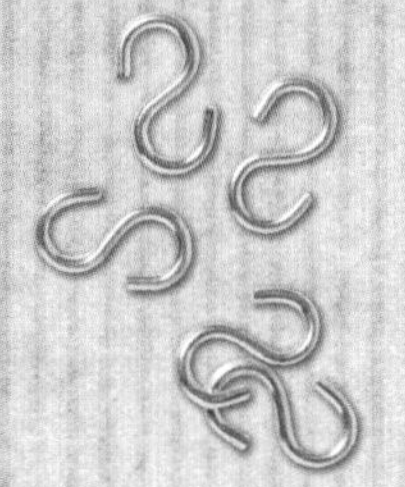

S-hooks Very useful hooks found in the hardware section of your local DIY store. Used for hanging things up.

Screw collars You'll find these at your DIY/ hardware store in the ventilation section. They are used to keep ductwork hoses together, but are great for holding hanging glass jars too.

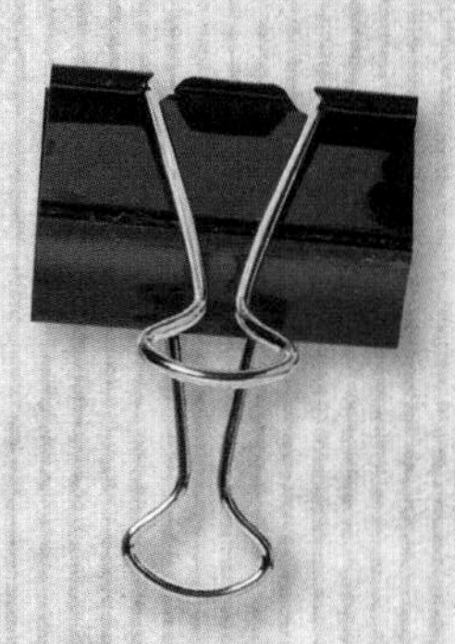

Binder clips Easily found at office supply stores, binder clips have lots of uses and come in a variety of sizes and colors.

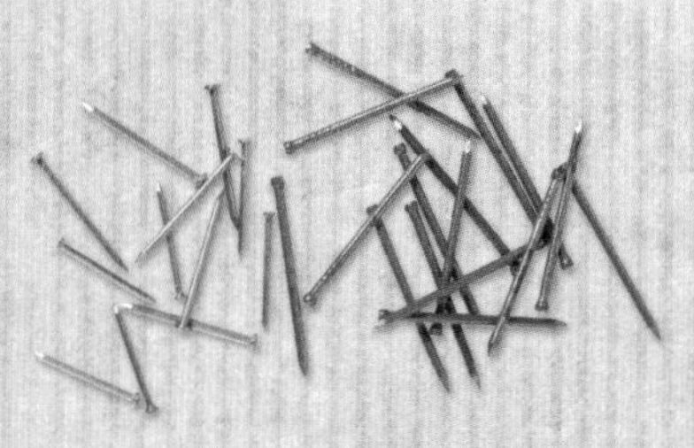

Nails You'll want a variety of the small nails often called brads. I use tiny ones that come with picture hanging kits. Little finishing nails are nice as they are smaller and aren't noticed as easily.

Wood screws If you want it to stay in place, use a wood screw because they really grip the wood tight. I suggest drilling a pilot hole and for extra durability add a little wood glue to the screw before setting it.

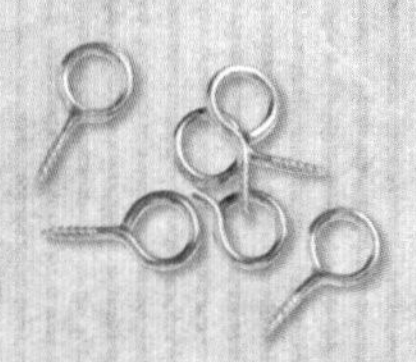

Small brass eye screws Add these to the back of pictures to take the picture-hanging wire.

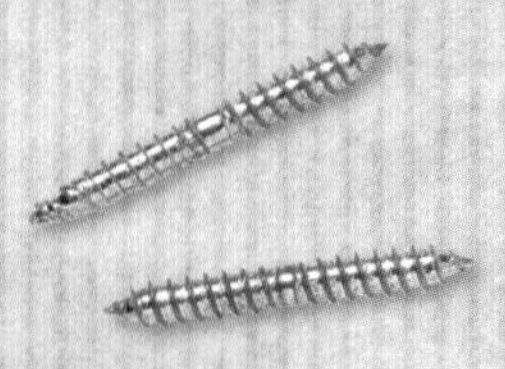

Dowel screws These are metal dowels with a screw tip on each end. They come in a variety of lengths and sizes and are found in your hardware or DIY store. When purchasing dowel screws I make sure I have a matching size drill bit, nuts and washers to go with them for a secure fit.

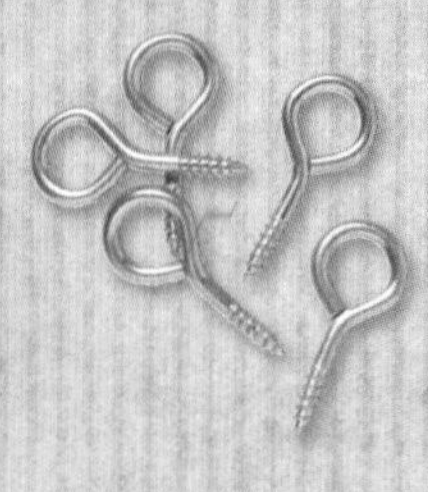

Eyelet screws or screw eyes I use these in a good majority of my woodworking projects because they are easy to insert—just drill a small pilot hole and screw them in—and they are durable. They also allow my project to fit flat against the wall.

Thumbtacks You can purchase already decorated thumbtacks, but they can be a little pricey and they may not match your décor. They are so easy to decorate yourself and can readily be sourced at home and office supply stores.

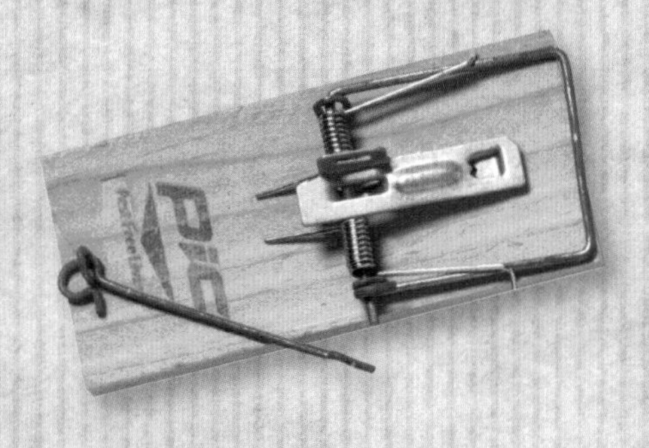

Mousetraps Not just for catching mice! These are great for sticky note holders. Train your eye and think creatively for other things you can repurpose and use in a new way.

CDs/DVDs Reuse old disks that don't work any more, or which hold material you no longer need—best to check what's on it first if it's a used disk!

Suction cup with hook An easy way to hang something on glass without marring the surface. They can easily be removed and relocated.

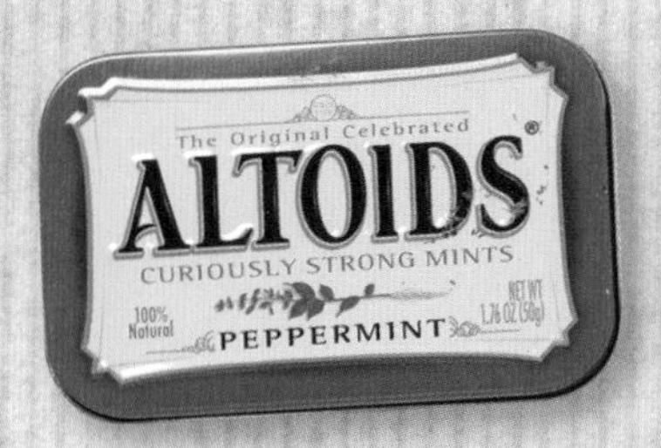

Mint tin Save all your mint tins (or similar small tins with hinged lids) as there are loads of things you can do with them. They are easy to paint, hold lots of small items, and can fit snugly into your purse or briefcase.

Notebook Mini notebooks are sold at office supply stores and are so handy for a variety of things. I upcycled a set into a flip calendar.

PVC pipe This comes in a variety of widths and lengths. I use ½ in. (1.25 cm) and 1 in. (2.5 cm) diameter. Many DIY stores will cut the pipe down for you.

Knobs I keep a steady supply of door and drawer knobs in all shapes and sizes on hand in my studio. They are great to create a pedestal base and hang things from. Some craft stores carry vintage looking knobs, but flea markets are a great place to find them as well.

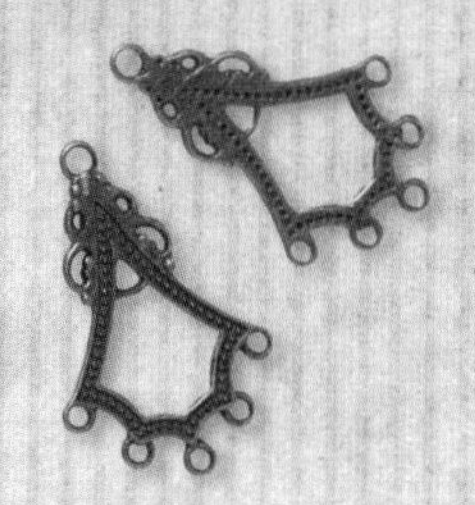

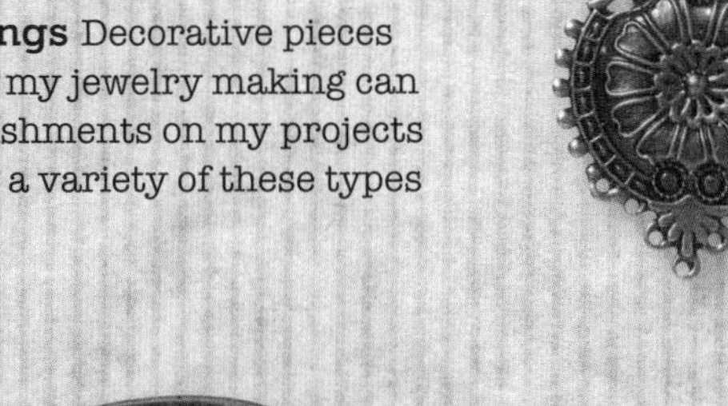

Metal jewelry findings Decorative pieces that I typically use for my jewelry making can also double for embellishments on my projects too. Craft stores carry a variety of these types of findings.

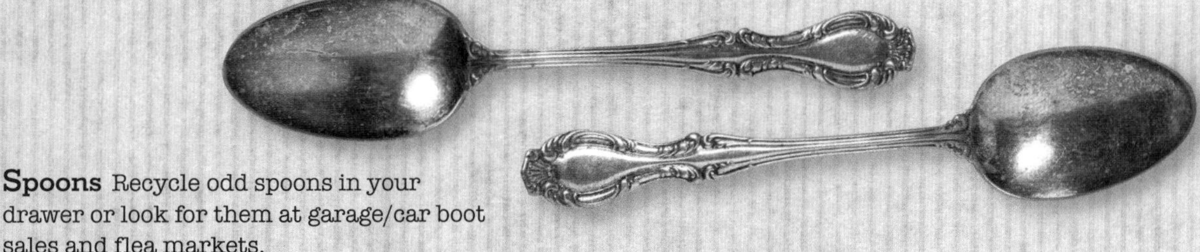

Spoons Recycle odd spoons in your drawer or look for them at garage/car boot sales and flea markets.

techniques

This section contains some basic skills and techniques that are used multiple times throughout the book so I've collected them here for easy reference. I recommend that you familiarize yourself with these techniques, because it will make it easier to complete the projects. Additional techniques specific to a particular project are detailed within that project.

Special finishes

I particularly like the look of vintage but sometimes it's not always easy to find items already distressed and aged. New items can be given an authentic shabby chic look by distressing, antiquing, or adding a crackle glaze finish. See pages 28–29 for recipes to make your own finishes.

Antiquing and aging

Combine approximately equal amounts of decoupage medium and acrylic paint to make a glaze. Brush or sponge the glaze onto the item to be antiqued. Remove excess glaze with a rag, leaving paint glaze behind in the cracks and crevasses if desired.

A Helpful Hint: Fabrics can be aged by dipping them in tea, squeezing out the excess water, and allowing them to dry. I do this a lot with muslin and linen as it gives them a beautiful vintage look.

Crackle glaze

This is a very easy way to get a weathered look, especially on wood. You need to work rather quickly when applying the top coat. I recommend practicing on a piece of scrap wood or other suitable material first.

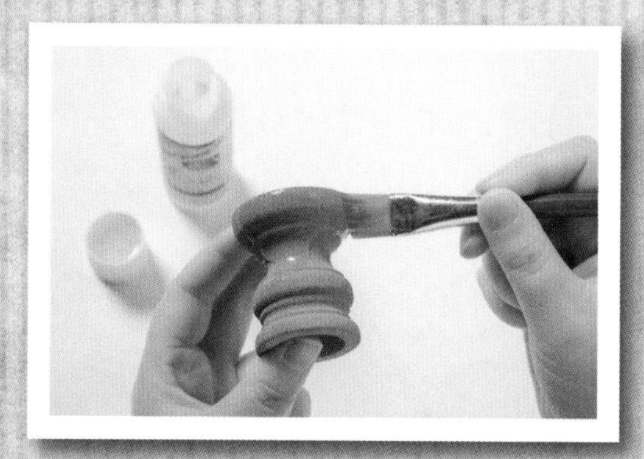

1 Paint the item with the base color. Apply the crackle medium and allow it to dry slightly—not completely.

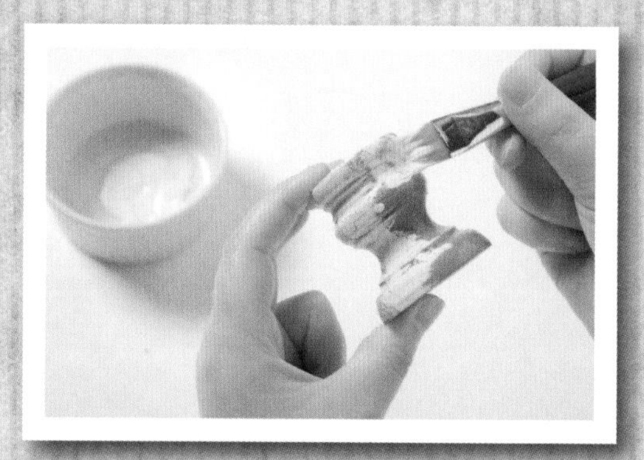

2 Apply the top color over the top of the crackle glaze with a brush, being careful to apply in one direction. As the top coat dries, the crackle finish will appear; the thicker the coat the larger the crackle—I like to have random areas of thick and thin. Experiment with this technique to attain the look that most suits your taste.

Distressing and adding visual texture

I almost always distress my surfaces in some way and these are a few of my favorite techniques.

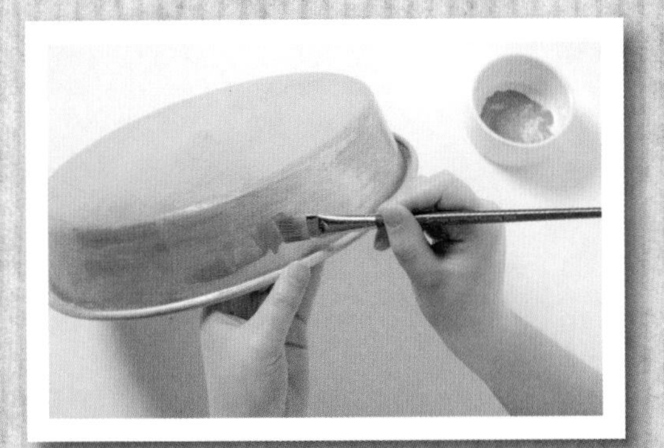

1 Sand the item if necessary to roughen the surface so the paint will stick. Paint the surface in the desired color and let dry. You can apply the paint as thick or thin as you like and also randomly missing some areas adds a nice effect.

2 Distress the surface slightly by rubbing gently with the sanding block to remove a little bit of the paint. Do this especially on any raised areas like corners and embellishments to give a natural-looking, worn appearance.

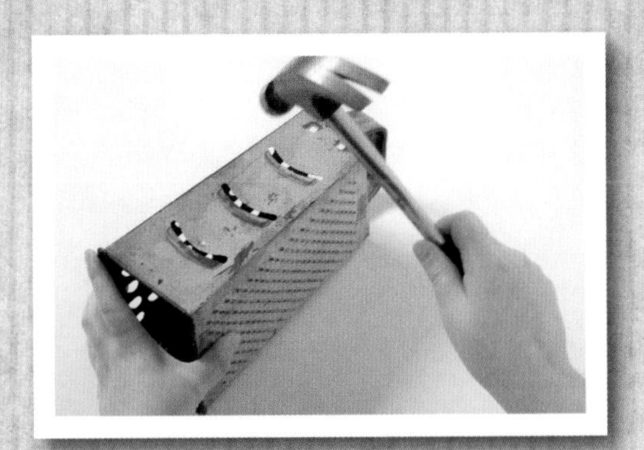

On painted metal, hammer over the surface at irregular intervals to distress the finish. You can also create a worn finish as above by sanding a little of a painted finish away. Hammering and sanding combine to create really nice results.

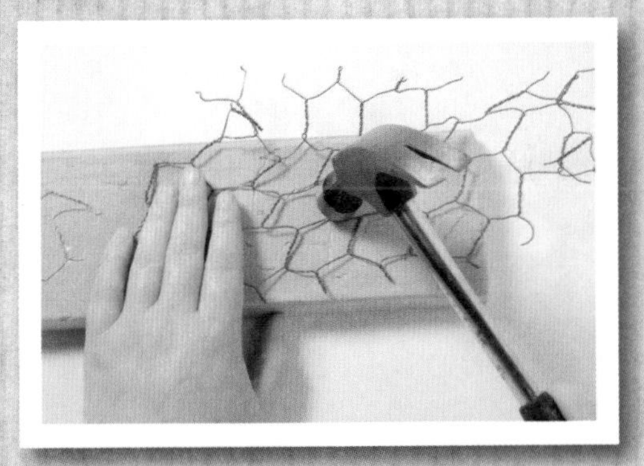

Paint wood in your desired color and then place a piece of chicken wire mesh over the top and hammer lightly to add some distress patterning to the surface. You can also distress with other objects such as nails, screws, or any other object with a suitable texture that will withstand the hammer. Experiment!

Adding decoration

These techniques can be used on almost any project, but find additional decoration techniques within the projects themselves.

Inking edges

Sand the edges all around so they are smooth. Shade the edges with a little color using a cosmetic sponge and a brown ink pad.

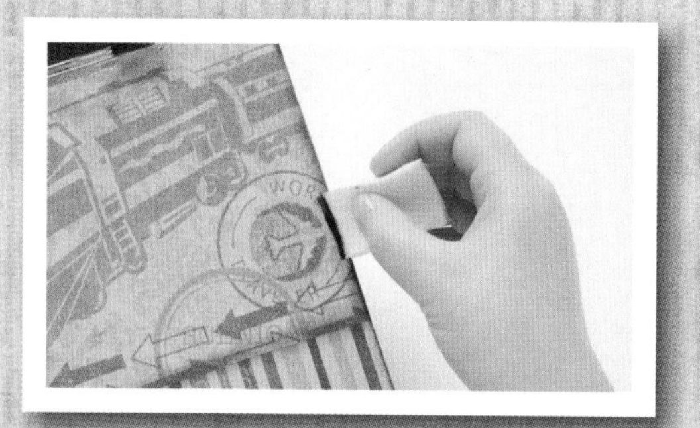

Decoupage

1 Paint the base item with a coat of decoupage medium, making sure it is all covered evenly. Do not allow it to dry—on larger items it may be necessary to work on small areas at a time.

2 Set paper down over the decoupage medium and then apply an additional coat of decoupage medium over the surface of the paper. Smooth out any bubbles. On larger objects, work from the center out toward the edges. If necessary, reinforce corners with a little extra medium. I decoupage a lot of my items because it provides a good water resistant finish.

Drilling and cutting

I've selected materials for this book that can be cut using hand tools. Caution—please take care when drilling and cutting to prevent injury.

Drilling holes

Before drilling into metal, mark the drilling position with a punch so the tip of the drill will not slip. Hold the drill as straight and upright as possible otherwise your hole will be at an angle.

When drilling into any small item make sure it is held securely in a vise to avoid accidents.

Cutting

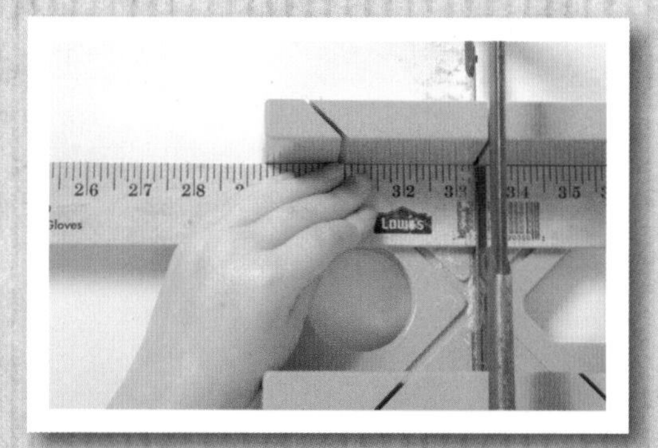

Wood is usually fairly easy to cut. Make sure the saw is sharp and push the blade smoothly—the cut stroke is generally when you are pushing rather than pulling back. Apply gentle yet firm pressure to the saw but let the saw do the work for you. This will save your arms!

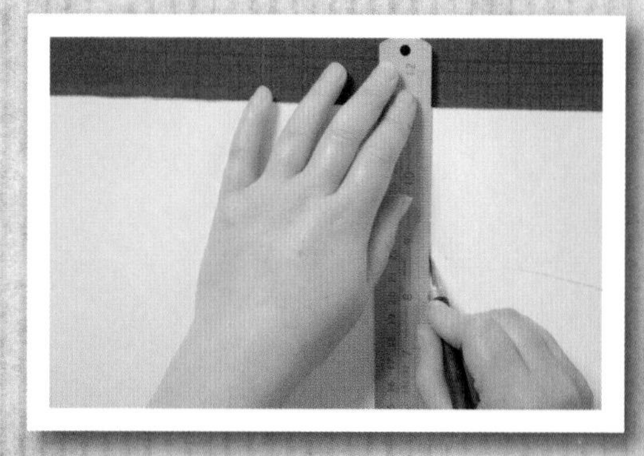

Cut foam board on a cutting mat to protect your work surface. Use a new blade in the craft knife and cut against a metal rule—do not use a plastic or wood rule because the edge is very easily damaged. Take care to keep your fingers away from the blade at all times.

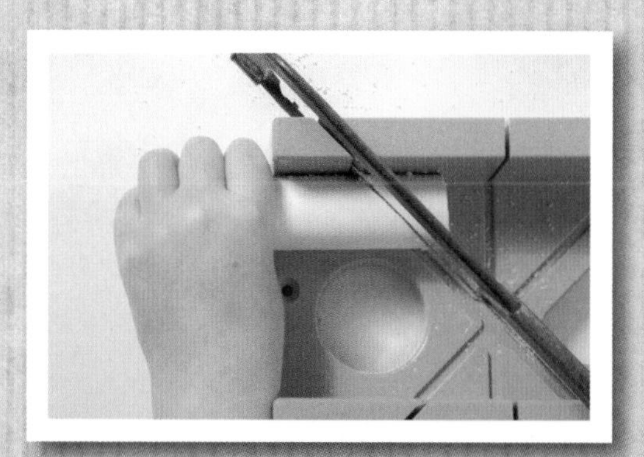

When cutting PVC tube—or any material—at an angle, use a miter box to be sure all the angles are exact. You can also use the miter box to get clean corner angles when making a frame.

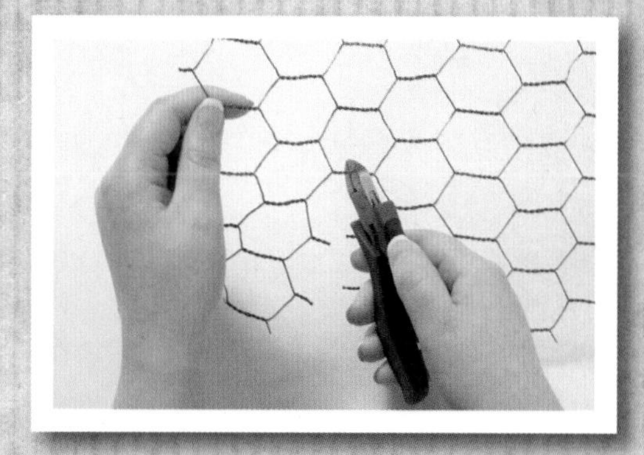

Cut chicken wire mesh roughly to size, then shape by snipping through the twisted wire join between the open holes.

Attaching and assembling

We use several ways of attaching items together, from simple gluing to drilling and screwing.

Using double stick

Double stick materials come in a sheet or on a roll in different widths. The sheet has a backing on both sides, while the tape is on a roll and only contains a backing on one side.

1 Remove one side of the backing. Apply the item to the sticky side of the tape.

2 Cut out around the item with sharp scissors if necessary—the cutting doesn't have to be perfectly neat around the edges at this stage.

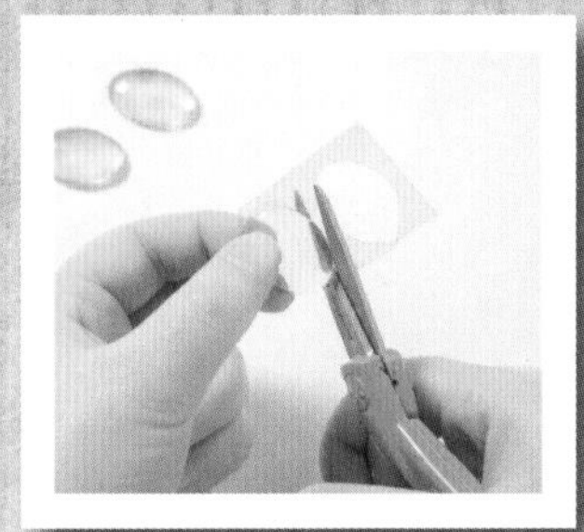

3 Remove the remaining backing sheet from the double stick tape on the item and apply to the background. Cut out if necessary.

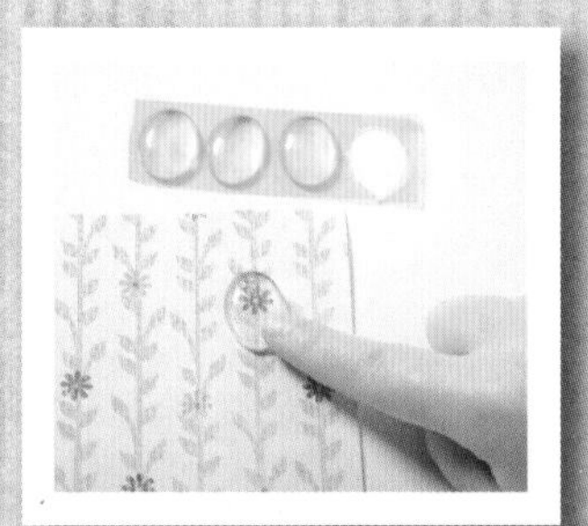

A Helpful Hint: I remove the white backing first and apply the item. If you heat the pink/red liner with a heat gun for a few seconds, this will cause the plastic liner to curl and make it easier to remove.

Opening and closing jump rings

Use this simple method to open and close jump rings so you can use them to attach items together.

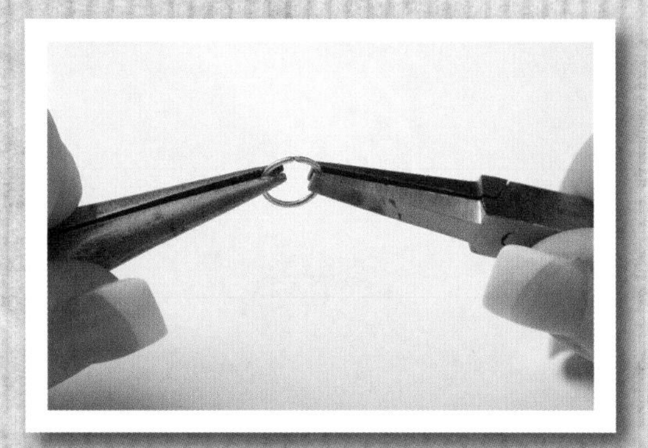

To open the jump ring, hold a pair of pliers on each side of the join and twist the pliers slightly in opposite directions to open up a gap.

To close the ring, repeat the twisting action in reverse to bring the two ends back together neatly.

A Helpful Hint: Don't open a jump ring by pulling the two ends apart, because this will distort the shape and it will be hard to get it back into a perfect circle. By opening the jump ring in the way shown here, it will stay perfectly round.

Attaching wire mesh

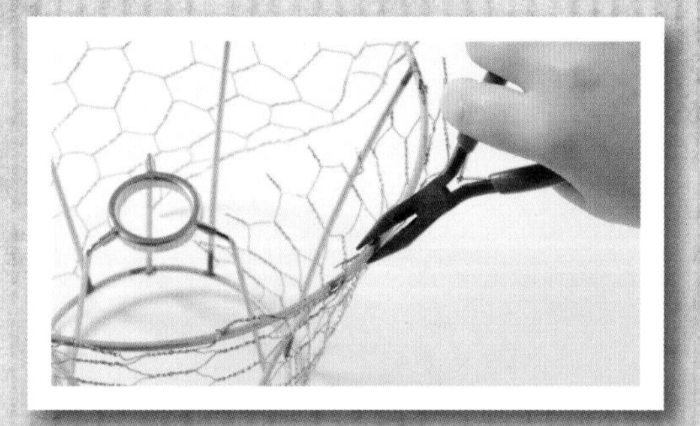

Cut the chicken wire to a length and height to fit the item roughly. Bend the ends around the frame, adjusting the chicken wire to fit and removing any excess with wire snips.

Painting and coloring

A few tips and techniques to help you achieve great results every time.

Preparing surfaces for paint

1 Lightly sand the item to be painted. On wood you are creating a smoother surface, while on shiny metal you need to create a slightly rough surface so the paint will be able to grip.

2 Paint the item as neatly as possible. Two light coats are better than one thick coat, which may chip. Let dry completely between coats.

Paint texturing

1 Apply one or more shades of acrylic paint to a piece of bubble wrap with a brush or sponge. Dab the bubble wrap onto the surface to create a sponged on look. Allow to dry completely.

2 For a more complex finish, paint the item with a base color and allow to dry. Using a cosmetic sponge apply a second color, letting a little base color show through. Allow to dry.

3 Sponge a third color lightly over the top of the second—again the base and second colors underneath should still show through a little to add even more interest.

Stamps

I like to look around the house for anything that will give an interesting pattern or texture.

Foam board I used left over foam board to create a rose stamp as you will see in the Magazine Bin project on page 126. Similarly you can use a piece from a foam plate to create the same kind of stamp.

Veggies Cut a shape or pattern into a potato, apply paint and stamp. Other fun veggies to stamp with include green peppers (these create beautiful flower shapes) and celery sticks for half moon shapes.

Craft foam Cut the foam into the desired stamp shape and glue onto an acrylic backing to make a re-usable stamp.

Flat erasers Carve them with a craft knife into any shape desired.

Do it yourself recipes

I always look for ways I can make my own supplies and thus save some money. A search on an online pegboard such as Pinterest will give you lots of results. Here are a few of my favorite recipes that I've found and use on a frequent basis.

Chalkboard paint

1 Add approx. ¼ teaspoon unsanded grout into ¼ cup (60 ml) acrylic paint in your chosen color.

2 Mix the two together until well blended. The paint should still flow and not be a stiff paste. If it becomes a paste, add a little more paint or thin it out with water to get a nice consistency.

Glaze

Combine equal parts of decoupage medium to acrylic paint. You can adjust this recipe even further by adding more paint to make the glaze more opaque or less paint to make it even more transparent. Additionally, there are clear glazes available from home DIY stores that can be used.

Glue dots

I use a liquid gel glue called Tack-it Over and Over—it is a re-positionable glue that, once dry, still remains a little tacky. Simply apply small dots of glue onto a sheet of waxed paper and allow to dry.

Spray color washes

Mix together 2 parts acrylic paint with 1 part water and pour into a mini spray bottle. This saves loads of money and you can have custom colors that match your décor available at your fingertips.

Alcohol ink

Mix approximately 1 part powder fabric dye to 2 parts alcohol. Place in a spray bottle or other bottle with a tight lid because the alcohol will evaporate. The saturation of color is a personal choice, so experiment to find the right blend for you. To create a more saturated color increase the powdered fabric dye. You can also experiment with other powder colorants such as fruit drink mixes. It's fun to experiment!

Decoupage medium

While I used a clear purchased glaze in this book, you can easily make your own decoupage paste similar to wallpaper paste—though it will not be as transparent. Combine flour, sugar, water, and white vinegar together to form a pancake batter consistency. Store in an airtight container—although the mixture won't keep long, so just make the amount you need each time and toss away any left over.

Air-dry modeling clay

Mix together 14 oz (400 g) baking soda (bicarbonate of soda), 3 oz (75 g) cornstarch (cornflour), 8 ½ oz (250 ml) water, and 1 tablespoon cooking oil to form a dough. If the dough is sticky add a bit more cornstarch; if too stiff add a small amount of water at a time. Wrap with cling film or plastic wrap and store in a zippered bag in a cool, dry place.

CONTRIBUTION
WORLD

repurposing and recycling

spring tree

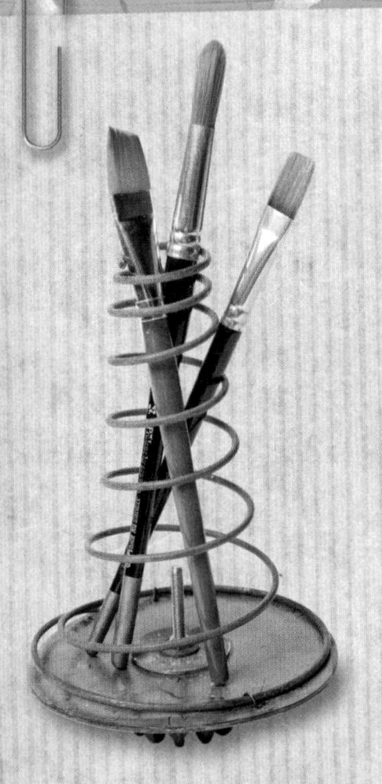

Originally I wanted to use an old floor brush to artfully display my paintbrushes—the tightly clustered bristles would have held the brushes upright—but I couldn't find a suitable brush. Then I happened on this old spring left over from another project. I loved the shape and after playing around for a while I came up with this spring tree brush holder! It's unique and a great conversation piece when friends stop by the studio.

- CD or DVD disk
- Acrylic paint in three colors
- Cosmetic sponges
- ⅝ in. (1.5 cm) metal washer
- Fender washer
- Bedspring coil
- Black marker
- Hole punch
- Faucet knob with bolt
- Black annealed steel wire
- Pliers

1 Sponge the disk with the base color and allow to dry. Sponge a second color over the top—don't try to make it too even, the idea is that a little of the base color should still show through.

2 Sponge a third color lightly over the top—again the two colors underneath should still show through a little. Also sponge all the colors onto the washers and the spring.

3 Align the spring centrally on the disk and mark the placement of holes to attach the fixing wire on each side.

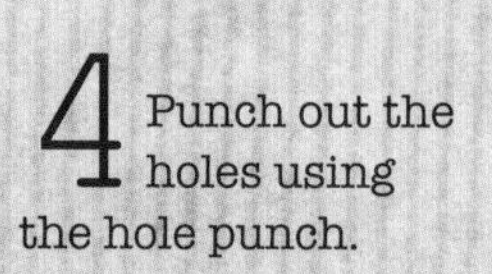

4 Punch out the holes using the hole punch.

just a note:

If you don't want to get into textured paint finishes, just choose a suitable decorative paper and decoupage (see page 24) it onto the disk and washers.

This spring tree can hold more than just brushes. Use it to hold your pens and pencils or even clip on small clips to hold notes and photos.

5 Thread the disk onto the bolt of the faucet knob through the center hole.

6 Add the small washer, which fits inside the center hole of the disk so it doesn't move around on the bolt.

7 Place the painted fender washer on top and then screw on the nut to hold everything secure. Cut short pieces of wire and feed up through the holes in the disk, over the bottom of the spring, and back through the next hole. Twist the wire underneath to hold the spring securely in place.

mousetrap clips

When it comes to creative organizing, thinking outside the box can be a fun way to add additional charm to your studio or home environment. A mousetrap provides a great way to display notes that you want to keep in front of you—attach a magnet to hang on the refrigerator or memo board, or add a pin back for a bulletin board. Your local scrapbook store is a good source for finding embellishments and decorative paper to make the clips unique and according to your décor.

- Mousetraps
- Pliers
- Decoupage medium
- Decorative paper
- Cosmetic sponge
- Sanding sponge
- Heavy-duty scissors
- Large metal jewelry finding
- Tacky glue
- Flower jewelry findings
- Buttons (optional)
- Ribbon (optional)

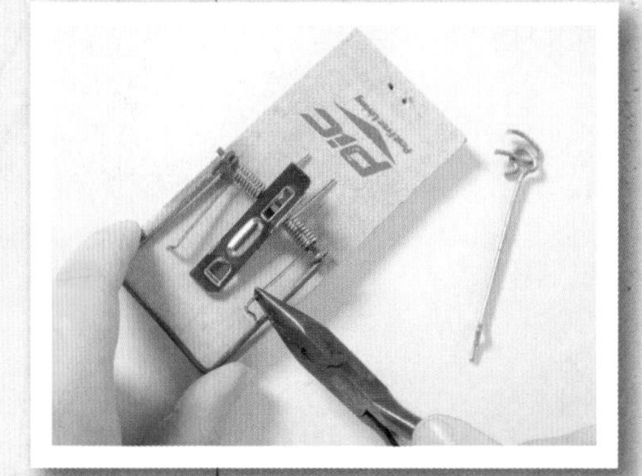

1 Remove the bar of the mousetrap assembly from the end, including the eye pin, and discard. Lift the spring bars off at each side to disarm the clip mechanism.

2 Unhook one end of the clip; the whole clip assembly will now slide off. Keep the clip and springs but discard the central metal plate.

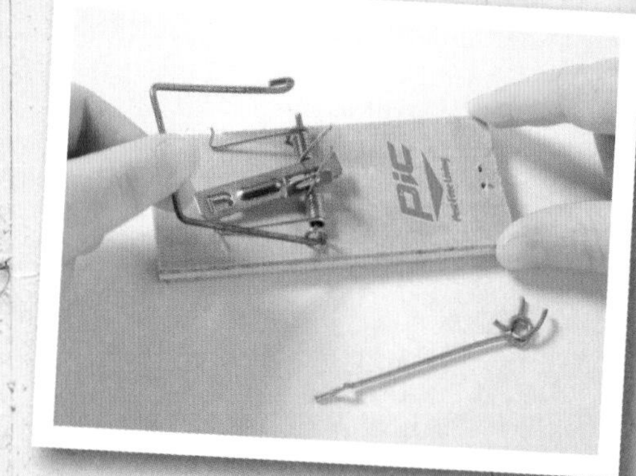

3 Decoupage the paper to both sides of the base (see page 24). Sand the edges smooth. Cut the finding in half using the scissors.

just a note:

Before you take the mousetrap apart, take careful note which way around the springs are on the clip, because you will need to reassemble the spring clip at the end. If you need to, snap a quick photo with your phone so you can see what the assembly looks like.

See page 29 for a recipe to make your own decoupage medium.

4 Glue one half of the finding on the reverse of the base so it sticks up over the edge and will show on the right side.

5 On the front side, reassemble the spring clip. Attach the flower finding and other embellishments of your choice with glue.

upcycled cigar box

Sometimes cigar boxes are rather frail because of their age—the lid came off this particular box despite the rest of it being in relatively good shape. I secured all the sides with glue and re-attached the top using lace as the hinge; I love that these little imperfections add to the charm of the box. These boxes are so handy to have on hand to store all sorts of things.

- 2 patterns of decorative paper
- Photographs printed on transparency
- Circular item as template
- Jewelry pendant blank
- Pencil
- Scissors
- Craft glue
- Two-part epoxy glue (optional)
- Cigar box
- Length of narrow ribbon
- Double stick tape
- Jewelry jump ring
- Pliers
- Spray adhesive

1 Use a photo editing program to print a photo onto transparency film. Find a suitable item to use as a template and mark and cut out circles from paper and transparency to match the size of the jewelry pendant blank.

just a note:

I used an Artisan X-plorer die cutting machine (see page 13) and corresponding dies as described on page 70 to make the circle cutting in this project quick and simple. Not to worry if you don't have one though—simply use a pencil and paper to trace a template for your circle, as shown, and you're good to go.

R. G. DUN
BOUQUET
R.G.DUN
TRAVELER

just a note:

You can use decoupage medium instead of spray adhesive to fix the paper to the box if you prefer. If you use spray adhesive, be sure to work in a well ventilated area.

This box can also be made as one of the internal fittings for the Crafts-on-the-go Trunk on page 98.

You can make dividers for this box by cutting lengths of wooden yardstick to size and glueing them in! I love things with tiny compartments—don't you?

2 Drop the paper circle and transparency into the jewelry blank—secure with a dab of glue in an inconspicuous place. You could also fill in the blank with a little two-part epoxy glue, following the manufacturer's instructions, to add a glass-like finish.

3 Measure and cut a piece of decorative paper to fit on top of the box. Cut two lengths of ribbon and tape the ends on the back of the paper so they cross in the center on the front as shown.

4 Open a jump ring (see page 26 for instructions on how to open a jump ring correctly) and hook it around the two strands of ribbon at the front.

5 Thread the pendant onto the jump ring and then close it again, making sure to align the ends neatly.

6 Apply spray adhesive to the back of the paper and attach it to the top of the box.

clock knob with pencil bucket

Hooks are great, but this decorative knob is just as useful and much prettier. I created the hanging bucket to hold all my pencils but it is the perfect size to hold all kinds of things like coloring books and crayons or even stuffed animals. Change the knob to fit your theme.

- 4 x 4 in. (10 x 10 cm) gesso board or wooden block
- Small metal bucket
- Paintbrush
- Acrylic paint in off white and gray, or colors of choice
- 2 x 2 in. (5 x 5 cm) decorative block
- 2 picture hangers
- Hammer
- Drill and ¼ in. (6 mm) drill bit
- Wood glue
- Knob with bolt
- Sanding block
- Ribbon in coordinating color/pattern

1 Paint the gesso board and bucket with gray acrylic paint, and allow to dry. Paint the decorative wood block with off white and allow to dry.

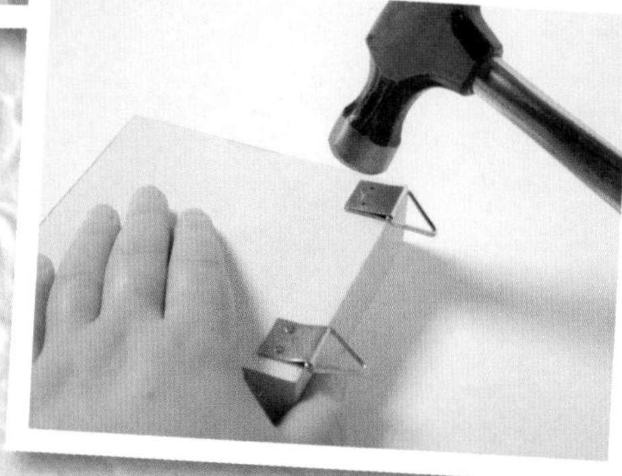

2 Attach the two picture hangers to the back of the gesso board.

3 Carefully drill a ¼ in. (6 mm) hole through the center of the decorative block.

CONTRIBUTION
MANUFACTU
-40-
Marc Vidal

just a note:

For a more contemporary look, add a simple knob and do not distress the bucket. For a kid's theme, use a suitable knob and paint the bucket with primary colors.

4 Glue the decorative block to the gesso board with wood glue. Add a little glue to the bolt of the knob and insert it into the center hole.

5 Paint the bucket with gray acrylic paint—keep the finish fairly rough for a vintage look. Allow to dry.

6 Use the sanding block to distress (see page 23) the surface slightly.

7 Knot the ends of the ribbon to the handles on each side to make a hanging loop. Hang the bucket on the knob and use to store pencils or whatever you like.

grate display

Give an old kitchen grater a new lease of life as a jewelry organizer. Don't worry if you can't find a vintage grater to use—the steps also show you how to distress a new grater to give it the vintage look.

- Square kitchen grater with a metal handle
- Gray and brown acrylic paint
- Paintbrush
- Sanding block
- Hammer
- Small block of wood
- Tape measure or rule
- Saw
- Drill and ¼ in. (6 mm) drill bit
- Crystal knob with bolt and matching nut
- Four finishing nails
- Two-part epoxy glue
- Vintage style metal door handle with back plate

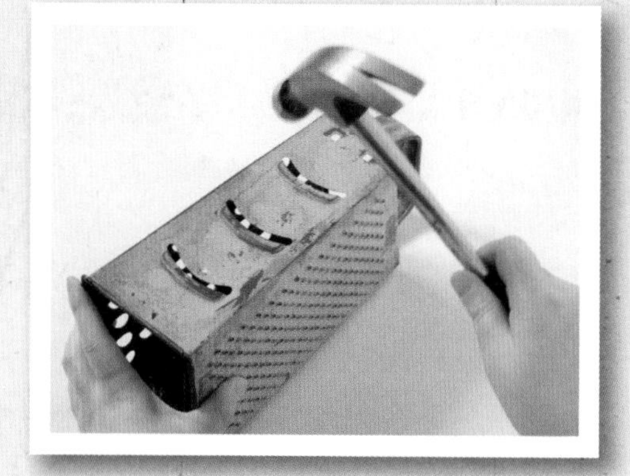

1 Paint all the surfaces of the grater with acrylic paint and allow to dry. Sand off some areas and hammer over the surface at irregular intervals to distress the finish.

2 Measure a block of wood to fit the inside of the base and paint it with acrylic paint. Drill a hole in the center and push the bolt of the knob through. Screw on the nut to secure the knob in place.

3 Push the block into the bottom of the grater so the crystal knob is on the outside. Hammer in a couple of finishing nails on either side of the grater to hold the wood block in place.

just a note:

You do not have to use exactly the same knobs as we have used here—look around hardware stores, yard sales, and thrift shops to see what you can find that would do the job just as well.

To keep the grater upright while the glue is drying, prop it up against a couple of tall drinking glasses, or something of similar height, and place in an area where it is not likely to get knocked over.

4 Put a thin layer of two-part epoxy glue around the top of the metal door handle and position the crystal knob on top. Leave to dry completely.

three-tier fountain storage

Inspired by the tiered shape of classical fountains, this is the ideal way to store small items in random sizes and shapes but at the same time have easy access to each item. I also like that it uses air space and takes clutter off my work surface.

- Set of cake pans, 6 in. (15 cm), 8 in. (20 cm), and 10 in. (25 cm) in diameter
- Sanding block
- Acrylic paint in desired colors (tan and light gray)
- Paintbrush
- Four short candlesticks
- Two tall candlesticks
- Crackle medium
- Antiquing glaze or wood stain
- Cosmetic sponge or small brush
- Drill and ¼ in. (6 mm) drill bit
- Two ¼ x 1½ in. (6 mm x 4 cm) fender washers
- Two 2 in. (5 cm) long wood screws
- ¼ in. (6 mm) dowel screw
- Silicone-based glue

1 Sand the outside of the cake pans to roughen the surface so the paint will stick. Paint the pans and candlesticks in the desired colors and let dry.

2 Distress the surface slightly by rubbing gently with the sanding block to remove a little bit of the paint.

3 Paint the small candlesticks with the base color. Apply the crackle medium and allow it to dry slightly. Apply the top color—the crackling will appear as the top coat dries.

4 Make up an antiquing glaze as described on page 22. Brush or sponge the glaze onto all the candlesticks.

5 Drill a hole in center of each cake pan. Drill a hole in the center of each tall candlestick at the top and the bottom.

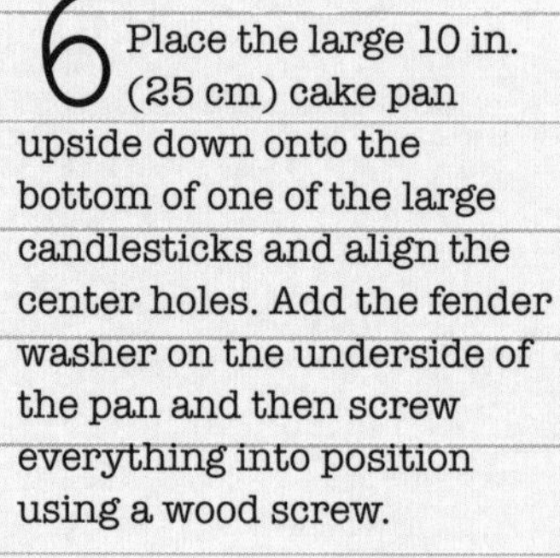

6 Place the large 10 in. (25 cm) cake pan upside down onto the bottom of one of the large candlesticks and align the center holes. Add the fender washer on the underside of the pan and then screw everything into position using a wood screw.

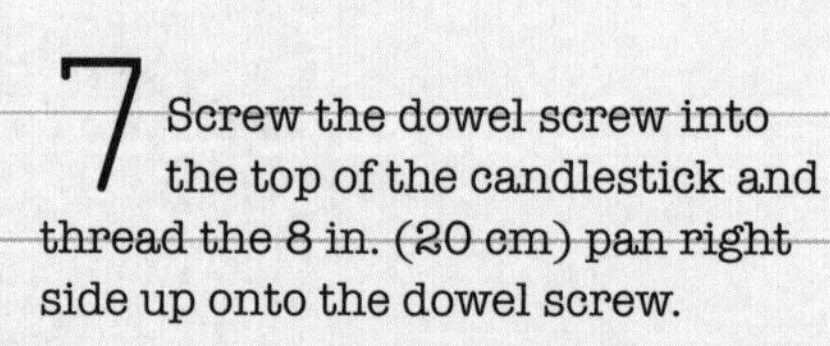

7 Screw the dowel screw into the top of the candlestick and thread the 8 in. (20 cm) pan right side up onto the dowel screw.

8 Thread the base of the second tall candlestick onto the top of the dowel screw and rotate to tighten.

9 Place the smallest pan, right side up, on top of the second candlestick, add the second fender washer and align all the center holes. Screw firmly into place with another wood screw. Turn the whole thing upside down and glue on the feet using a silicone-based glue.

just a note:

To make the paint stick to the surface of the metal—or any other slick surface—paint a base coat of tan chalkboard paint (see how to make your own chalkboard paint on page 28). The paint has a slight grit to it, allowing the top coat to stick to the surface.

This same project idea can be used to create cake stands and cupcake holders. Try substituting decorative plates for the pans or follow the instructions using colanders instead of pans.

To screw on the feet, drill four holes in the 10 in. (25 cm) pan at the 2, 4, 8, and 10 o'clock positions in step 5, and drill into the bottom of the four small candlesticks. Attach the feet using four long wood screws.

Instead of making up your own antiquing glaze you can use a little wood stain in your chosen color—although this will take longer to dry.

lamp note post

Recycle an old lampshade into a neat stand on which you can use small clothes pins to post pictures, notes, and other small papers. This would also be cute placed inside a shallow bin that holds paper clips, binder clips, and other office supplies.

- Lampshade with a metal wire frame
- Scissors
- Mineral spirits or lighter fluid (optional)
- Chicken wire
- Wire cutters
- Protective gloves (optional)
- Pliers
- Printed pages from a dictionary
- Decoupage medium
- Cosmetic sponge
- Tall and short wooden candlesticks
- Drill and ¼ in. (6 mm) drill bit
- Thick cardboard
- White tacky glue
- Wood glue
- Decorative knob
- Small clothes pins
- ¼ in. (6 mm) dowel screw
- Silicone-based glue

1 Carefully remove the old fabric from the lampshade if necessary. Remove any old adhesive from the frame with a little mineral spirit or lighter fluid.

2 Cut the chicken wire to a length and height to fit the lampshade roughly. Bend the ends around the metal lampshade rings, adjusting the chicken wire to fit.

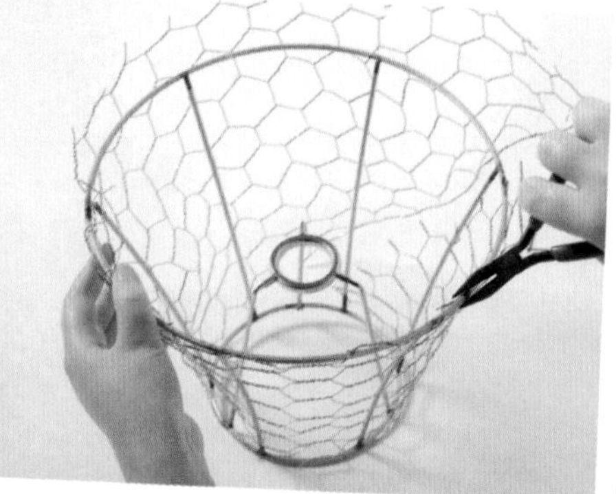

3 Decoupage dictionary paper onto each of the candlesticks (see page 24). Drill a hole into the bottom of the short candlestick.

just a note:
You need a shaped lampshade with metal struts between the top and bottom rings as a base.

4 Cut a circle of cardboard to fit the center lamp-holder ring, decoupage with some dictionary paper and glue in place. Leave to dry thoroughly and then drill a hole in the center. Apply a daub of wood glue into the hole in the short candlestick. Thread the screw of the decorative knob through the two holes to join the lampshade to the candlestick. Set aside to dry completely.

5 Glue the two candlesticks together with wood glue and let dry. Use clothes pins to attach notes, photos, and small items to the mesh at random.

jewelry carafe

Upcycle an empty bottle into an eye-catching storage place for all your small items of jewelry. Look out for interesting bottle shapes for this project—I rescued a soy sauce bottle that would have otherwise found a home in the landfill.

- Decorative metal candleholder
- Shaped small bottle
- Silicone-based glass glue
- Crushed shells
- Piece of paper

1 Press in the metal leaves on the metal candleholder as necessary so they fit together snugly around the base of the chosen bottle.

2 Add a thin layer of glue to the base of the bottle and push it down into the candleholder. Leave to dry completely before moving.

3 Carefully pour some crushed shells into the bottom of the bottle until it is about one third full. To avoid the filling material spilling, fold a square of paper in half and use it as a funnel. The crushed shells look nice while providing some weight so that your stand will not tip over easily.

just a note:

Prepare the bottle by removing any labeling and glue. Look at the bottle for its shape rather than what it is.

If you find the words or label hard to remove, you can always paint the bottle to disguise them.

You can fill the bottle with any kind of dry material to achieve different looks—try colorful beads, dried seeds or beans, or aromatic spices.

Look for something to make the metal base for this project in hardware or electrical stores—decorative parts from an old-style light fitting may well be the right sort of shape.

4 Hang your bracelets around the neck of the bottle, earrings around the lip, and place rings on the leaves of the candleholder.

hanging gutter caddy

The great thing about this preformed gutter is that you can easily cut it to any length you need and buy end sections that just snap on to instantly create a useful container. I've glued the end pieces on for extra security. You can hang as many of these in a line as you choose, or just make single caddies to stand on a shelf.

- Two 12 in. (30 cm) lengths of gutter
- 4 end pieces
- Acrylic paint
- Paintbrush
- Decorative paper
- Double stick tape
- Drill and 3/16 in. (4.5 mm) drill bit
- Two 12 in. (30 cm) wooden rules
- Glaze or wood stain
- Cosmetic sponge
- 2 screw eyes
- 8 small S-hooks
- 2 approx. 3 in. (7.5 cm) lengths of jack chain
- 2 approx. 8 in. (20 cm) lengths of jack chain

1 Paint the inside of all gutter and end pieces and cover the outsides with decorative paper using double stick tape. Distress if desired (see page 23). Drill a hole on either side of one pair of end pieces near the top edge and one hole only at the top edge near the back on the second pair of end pieces. Glue the end pieces with one hole onto one length of gutter for the top caddy, and then drill a hole through the bottom of the gutter and through the bottom lip of the end piece at each end.

2 Glue the end pieces with two holes onto the other length of gutter for the bottom caddy. Color the two lengths of rule with glaze or wood stain to give it an antique look (see page 22). Glue a rule onto the front top edge of each caddy. Turn the top caddy over and screw one of the screw eyes into each hole in the bottom of the end pieces.

just a note:

In step 1, drill the single hole in the end pieces of the top caddy near the back of each end piece, and in step 3 position the 8 in. (20 cm) loop of chain so the back section of the chain is shorter than the front. This allows the caddies to hang at an angle against the wall so you can see the contents easily.

Only the top caddy has holes drilled through the bottom at each end to take the screw eye for the hanging chains—do not drill the bottom of the base caddy.

You can just make these caddies to sit on a shelf instead, in which case you do not need to do any drilling for the hanging holes and will not need the screw eyes, S-hooks, or chain.

3 Attach an S-hook to each top hole in the end pieces of the upper caddy and add a 3 in. (7.5 cm) length of chain to each hook. Attach an S-hook to each screw eye on the bottom of the upper caddy. On the lower caddy, attach an S-hook into the each top hole in the end pieces and add the ends of an 8 in. (20 cm) chain to each pair of hooks. Attach a link on each chain to the S-hooks on the bottom of the upper caddy. Hang the caddies by hooking the short chains onto two nails or hooks on the wall.

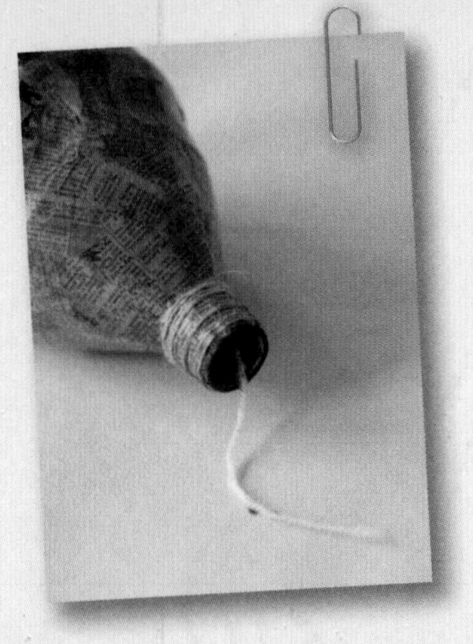

yarn bin funnel

Yarn and twine are always unraveling and getting mixed up, but this yarn bin will keep them neat and give you easy access too—thread the yarn end through the handy spout at the bottom and then just pull out what you need and snip off.

- 2-liter soda bottle with top shaped like a funnel
- Masking tape
- Scissors
- Dictionary paper
- Ink in color of choice
- Cosmetic sponges or paintbrushes
- Decoupage medium
- Brown antiquing glaze
- Hole punch
- Wire cutters
- 22-gauge wire
- Pliers
- Twine
- 7-8 buttons in various sizes and coordinating colors
- Hot glue and glue gun

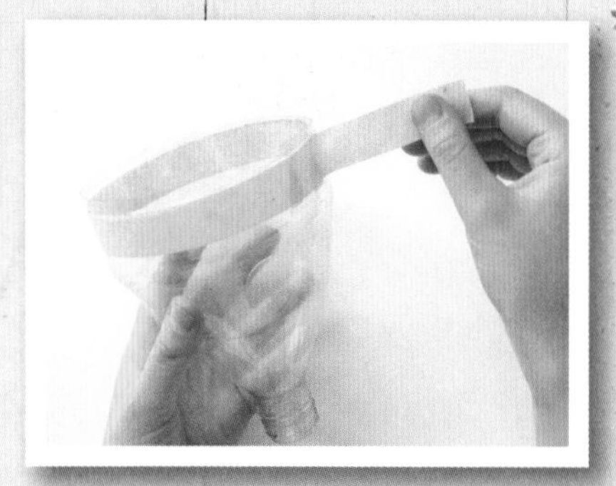

1 Decide on the depth that you want the funnel to be and wrap a strip of masking tape around the bottle to give you guide to cut a straight line. My funnel is about 5½ in. (14 cm) deep. Cut off the bottom part of the bottle along the edge of the tape.

2 Color the paper by applying ink directly onto it with a brush or cosmetic sponge. Tear the paper into strips and decoupage them (see page 24) to cover the entire outside and inside of the funnel. Allow to dry. Apply an antiquing stain (see page 22) and allow to dry.

3 Punch a small hole on either side of funnel near the top edge, positioned slightly towards the back with the holes about 3½ in. (9 cm) apart. Cut around 10 in. (25 cm) of wire and attach one end by threading it through a hole, bending the end upward and looping it around the main length. Repeat to add the other end of the handle on the other side. Closely wrap twine around the wire, securing the ends with a daub of hot glue. Add a small bow in twine to the handle and a tiny button. Wrap twine around the spout as an accent feature, if desired.

4 Attach a row of buttons around the front top edge of the funnel between the handles, using dabs of glue from the hot glue. Allow to dry thoroughly before using.

rule rack

Old yardsticks have a great deal of vintage charm; get the same effect by antiquing a new version from a DIY store. For this cool kitchen tool rack you could also use up segments from an old broken yardstick.

- Approx. 4 in. (10 cm) of wooden yardstick
- Small hacksaw
- Brown color wash
- Cosmetic sponge
- 2 brass screw eyes
- 12 in. (30 cm) length wooden dowel
- 2 vintage wooden thread spools
- Wood glue
- 12 in. (30 cm) wooden rule
- 2 decorative vintage teaspoons
- Pliers
- Drill and ¼ in. (6 mm) bit
- 2 wood screws
- Two lengths of wire

1 Cut two short lengths of the wooden yardstick each about 2 in. (5 cm) long. Antique the pieces of rule using brown color wash applied with a cosmetic sponge. Add a screw eye into the center top edge of each small piece. Color the dowel rod to match the pieces of rule.

2 Glue each end of the 12 in. (30 cm) long rule to one of the small yardstick pieces. Glue a wooden spool on top of the rule.

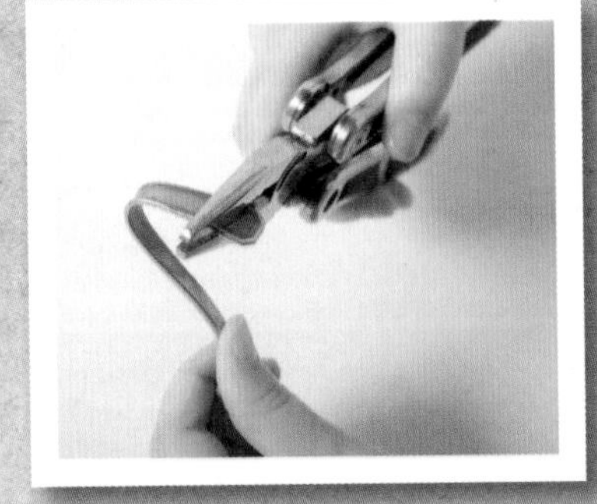

3 Use pliers to bend the handle of each spoon over into a hook. Drill a hole through the center of each spoon bowl, twist in a wood screw and glue to secure.

just a note:

We made a color wash with brown acrylic paint diluted with water, but you can use a wood stain—although it might take longer to dry.

Placing the spoon in a vise will help it bend evenly. Be careful as you shape the spoon because if you force it too much, it may break.

4 Push the screw on each spoon into the hole in the center of the wooden spool and glue in place. Prop this assembly up until it dries.

5 Place the length of wooden dowel into the hook on each spoon and wrap with wire to secure in place.

recycled cans

Why spend big money on an ordinary organizer? Make good use of all those empty veggie cans by turning them into storage pots—and you could set them on a lazy Susan so you can access what you need easily.

- Empty tin/aluminum cans
- Acrylic paint
- Paintbrush
- Printed paper, such as pages from a book
- Scissors
- Pieces of muslin
- Embellishments, such as lace motifs and metal findings
- General-purpose glue

1 Remove the paper labels from the cans. Paint the can in your desired color using acrylic paint. Cut a wide strip of printed paper to wrap around the can and glue it in place.

2 Tear a strip of fabric long enough to go around the can a couple of times with extra to tie. Wrap the fabric around the can and knot the ends—you can also glue the strip in position for extra security.

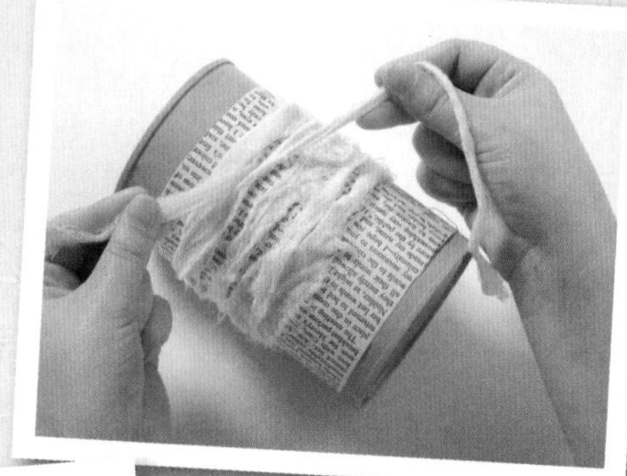

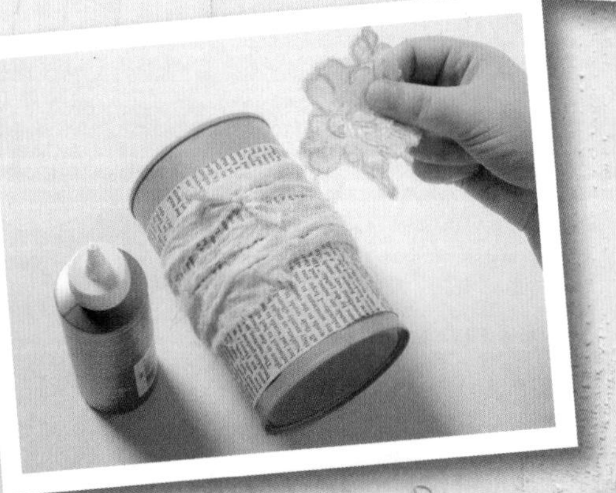

3 Add embellishments on top of the fabric strip to hide the knots, such as motifs cut from lace or a small crochet doily.

4 For extra decoration, glue on a small decorative metal jewelry finding. Try out other options, such as threading small items onto twine knotted around the can.

just a note:

If the original label has left glue on the can, remove it with a little mineral spirit or lighter fluid.

For alternative embellishments you can use bits of broken jewelry, attractive buttons, small pieces of lace, paper doilies, old keys—the options are endless.

Decorate several cans in different sizes and place them onto a purchased lazy Susan.

WORLD

miniature clipboard

- Small clipboard
- Pencil
- Decorative paper in two designs
- Scissors
- Spray adhesive
- Craft knife
- Sanding block
- Cosmetic sponge
- Brown ink pad
- Tacky glue
- Hook-and-loop tape (optional)

As a stand-alone project this mini clipboard is ideal for carrying around in your purse to make notes. But it can also be made as one of the internal fittings for the Crafts-on-the-go Trunk on page 98.

1 Draw around the clipboard onto one of the pieces of decorative paper and cut the paper to size. Spray the back of the paper with adhesive and apply it to the front of the clipboard, cutting out neatly around the clip.

2 Cut a wide strip of the second decorative paper to fit across the lower part of the clipboard. Spray the back of the paper with adhesive and apply it to the front of the clipboard at the bottom. Sand the edges so that they are smooth and flush with the edge of the clipboard.

3 Shade the edges with a little color using a cosmetic sponge and the brown ink pad (see inking edges technique on page 24).

just a note:

If you will be using the clipboard as one of the internal fittings of the Crafts-on-the-go Trunk (see page 98), attach a piece of hook-and-loop tape with tacky glue to the reverse of the clipboard.

chapter 3

storage and display

artful bowls

So quick and easy to make, and you can add any kind of texture you choose to create a different look. Use these bowls to store small items of jewelry, coins, or keys. If you can't find a suitable bowl mold, you can also use a balloon instead.

- White air-dry clay, such as Cloud Clay
- Doilies for impression
- Small plastic bowl
- Decoupage medium
- Cosmetic sponge
- Rolling pin (optional)

1 Roll a large ball of clay in your hands and begin shaping with your thumb to form a bowl shape.

2 Press a doily inside the bowl to texturize the surface. If your texture didn't turn out right the first time, simply re-roll the clay and try again.

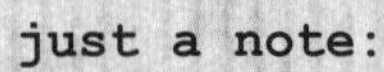

just a note:

Make your own air-dry clay following the instructions on page 29.

If you live in a place with high humidity, you may have to allow extra time for the bowl to dry.

If you want, you can paint the bowl using acrylic paint—allow to dry before sealing with decoupage medium.

Keep unused clay under wraps to prevent it from drying out. Have a little bowl of water handy to help moisten the clay if you need to.

Look around for anything that you think would lend an interesting texture to your bowls—maybe a grouping of paper clips, or perhaps a texture stamp.

3 Place the bowl over an upside-down small plastic bowl and form it over the bowl. Let dry according to the manufacturer's instructions. When fully dry, remove the bowl from the form and seal with decoupage medium. Let it dry completely.

artwork display frame

I used my die cutting machine and templates to create this project, but you can use it as inspiration to create frames in any shape you wish. This is a great project to get kids involved—get them to create their own special frame for their artwork. It's great for parents, too, because it allows kids to showcase their talent without taking up a lot of space.

- Artwork
- Frame template die
- Artisan X-plorer™ die cutting machine
- Plain paper or very thin cardstock
- Acetate
- Decorative paper
- Double stick tape
- Miniature hanging Hercules clips (optional)
- Twine (optional)
- Suction cups (optional)

1 Resize the artwork in your photo-editing program to fit the appropriate size frame, if necessary. Place the copy of the artwork on the cutting board of the die cutting machine, add the shaped frame template die on top with cutting side down, and then add the base board.

2 Run the stacked layers through the die cutting machine, which will cut out the picture to shape. Cut out a plain backing piece the same size and a shaped piece of acetate for the "glass."

just a note:

Keep the original artwork tucked away for safekeeping!

Hang the framed artwork with Hercules clips on a length of wire or twine or attach to the window using suction cups.

If you don't have a die cutting machine use the template on page 156. Alternatively, you could make up a shaped frame on your computer and use it as a template to cut the backing, artwork, acetate, and frame.

Look also for interesting shaped objects to draw around to make the frame, such as circles, diamonds, shaped rectangles, or hearts.

3 Use a pair of shaped frame nesting template dies, one inside the other, to cut a frame for the picture from decorative paper.

4 Stack the backing piece, acetate, and frame and glue together around the sides and bottom with a little double stick tape. Leave the top open. Trim the artwork if needed and slip the artwork into place inside the frame.

just a note:

The vintage ladies can be downloaded from the Internet—see suppliers on page 154. If needed, you can adjust the size of the clip art in your photo-editing program.

lace dollies

These are a wonderfully decorative way to both store and display lengths of pretty lace or ribbon. Display in glass jars or create the Wire Basket to hold them, as I have (see page 74).

- 2 vintage lady clip art prints
- Colored pencils
- Chipboard/thick cardboard
- Craft knife
- Decorative paper such as dictionary pages
- Decoupage medium
- Cosmetic sponge
- Paint stick
- Sanding sponge
- Glue
- Short lengths of lace or ribbon
- Glass-headed pins (optional)

1 Color in the vintage lady prints with colored pencils. Decoupage (see page 24) the ladies onto the chipboard and then cut out around the shapes with a sharp craft knife.

2 Decoupage dictionary pages to the paint stick. Cut the paint stick in half and sand any rough edges. Attach a paper doll to the top of each paint stick with glue.

3 Wrap a short length of lace or ribbon around each paint stick at the base of the doll and glue in place to finish the seam. Store lengths of ribbon or lace on the stick, secured with a glass-headed pin.

wire basket

We all know that baskets are a great way to organize similar items—but you may not know how easy it is to make your own. Why limit yourself to the sizes that are offered in stores if you only need something small, or something to fit in a particular odd sized area? Chicken wire is readily sourced at your home DIY or farm supply store; it's inexpensive and easy to mold into a unique form.

- Chicken wire
- Wire cutters
- Protective gloves
- Pliers
- Length of wide organza ribbon

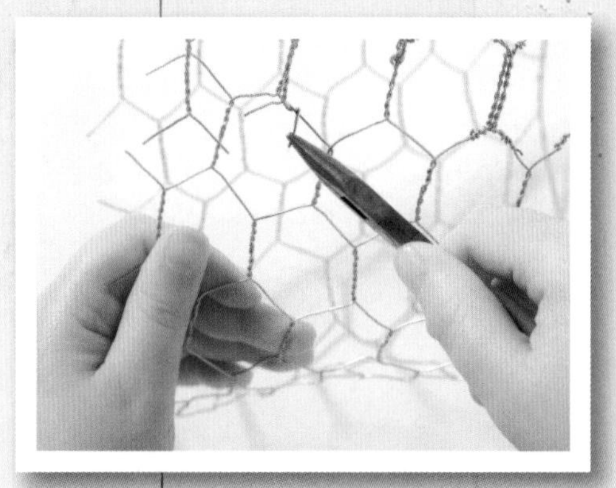

1 Decide on the height and diameter of the basket and cut a length of chicken wire mesh to match. Fold the ends around to meet, then twist the wire ends around each other to secure. To make it easier, match the pattern if possible.

2 Turn the basket upside down and push the bottom edges inward (you may wish to wear protective gloves). Twist the wire ends around each other in the middle to secure as before.

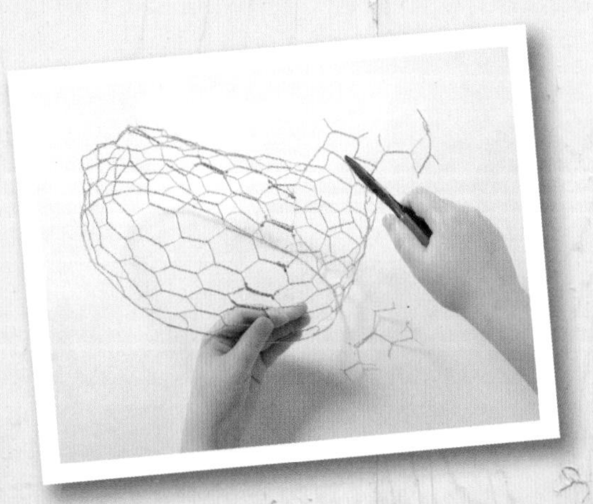

3 Continue to secure the bottom ends of the wire, clipping and trimming excess wire as required. Adjust the shape as needed with your hands.

just a note:

I left my basket in its natural silver color, but for a more contemporary look you can spray paint it in your desired color.

As you've probably already noticed, so many of the techniques in this book can be mixed and matched to customize. The basket liner project would work really well with this project too.

Caution–the cut ends of chicken wire can be very sharp. To prevent injury, wear protective gloves and be careful.

4 To soften the effect, wrap the length of organza ribbon around the basket and tie in a big bow at the front.

mason jar sewing kit

Keep your sewing supplies together neatly but still fully visible by storing them in this specially designed glass Mason jar with its pincushion lid. The pincushion is the ideal place for odd pins so they are fast to find when you need them.

- Mason jar with seal and ring
- Acrylic paint
- Paintbrushes
- Pencil
- Cereal box cardboard
- Scissors
- Double stick tape
- Small piece of fabric
- Fiberfill
- Tacky glue
- Twine
- Chalkboard paint
- Toothpick

1 Take the Mason jar lid apart and paint the ring. Use the seal as a template to draw a circle onto cardboard. Cut out the cardboard circle and put double stick tape around the edges. Remove the backing from the double stick tape ready for step 2.

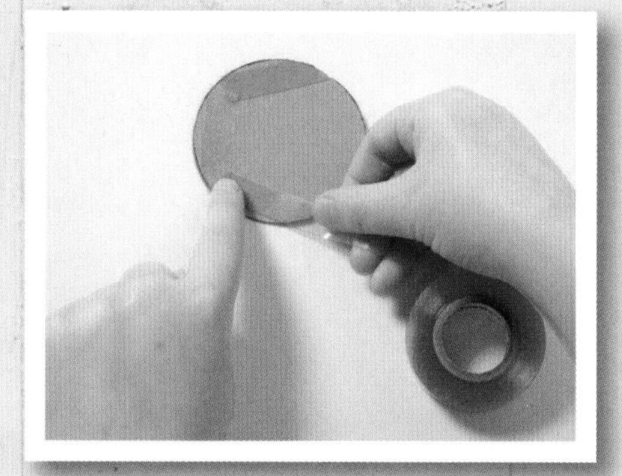

2 Use the seal as a template again to cut a circle from the fabric the same size and a second circle that is ¾ in. (2 cm) larger all around. Turn the larger circle wrong side up, place a handful of fiberfill in the center and add the cardboard circle on top with tape side upward. Fold the edges of the fabric over and stick them to the tape.

3 Run a narrow line of tacky glue around the underside of the lid ring on the inside—be careful not to get any glue on the outside. Carefully push the cushion into the lid so it makes a nice rounded shape in the center of the lid. Apply a thin layer of glue over the underside of the cushion.

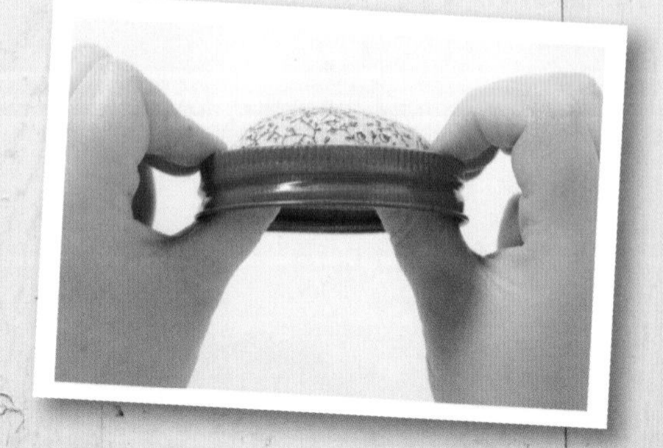

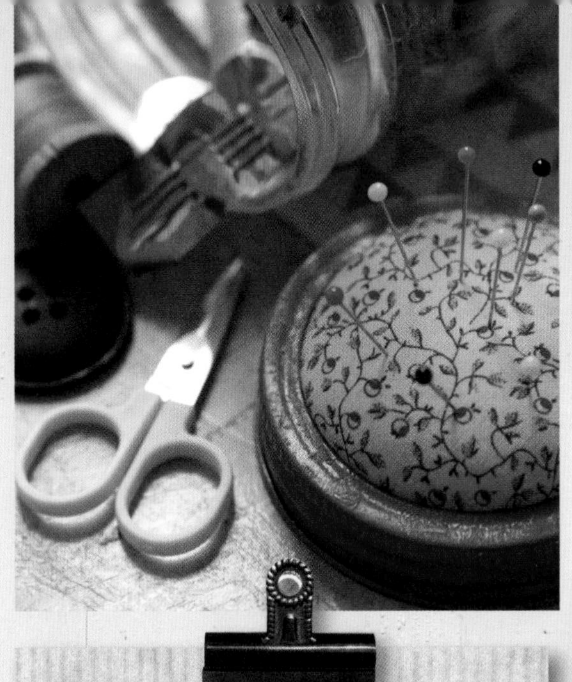

just a note:

For a vintage look, use old quilt scraps or recycle fabric from a shirt for the pincushion.

Label the jar using chalk, so it's easy to update when the contents change. Make several similar jars and create a grouping.

4 Add the smaller fabric circle inside the lid to neaten it. Add sewing supplies to the jar and replace the top. Wrap the top with a length of twine and tie in bow.

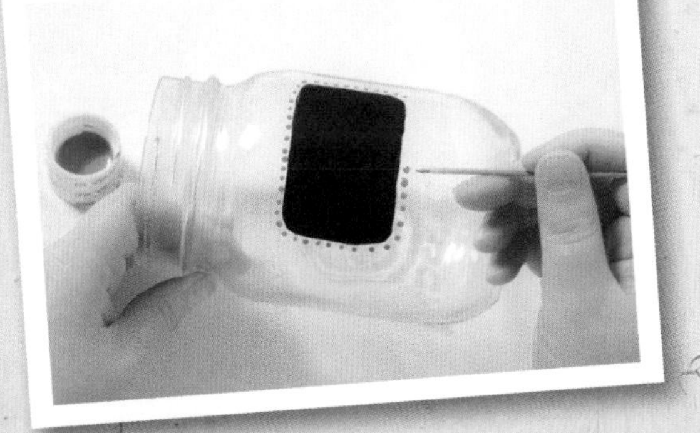

5 Paint a rectangle of chalkboard paint (see page 28) for the label on the front of the jar, adding several coats if needed. Let dry. Add any extra decoration you choose—we added a line of red dots in acrylic paint using a toothpick, to outline the label.

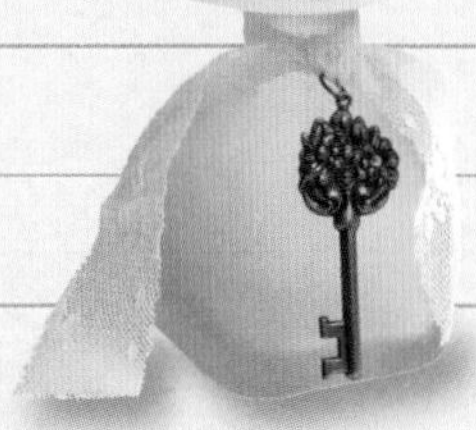

pedestal plate

This little dish is a decorative way to display hand soaps in your bathroom or as a tray for jewelry, or any of those little bits and pieces that scatter around and clutter the bedroom or bathroom. This is a very simple project with a big effect! Look around flea markets and charity shops for inexpensive yet decorative dishware. Or perhaps you only have one dish left of a favorite pattern—upcycle it and give it new life.

- Decorative dish
- Decorative bottle
- Short length of ribbon
- Decorative key, charm, or button
- Jump ring
- Pliers
- Silicone-based glass glue

1 Clean the dish and bottle thoroughly and remove any stickers or oily residue. Tie the ribbon around the neck of the bottle.

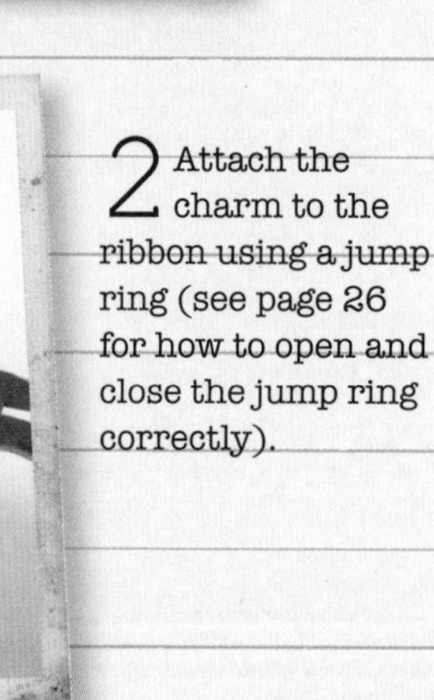

2 Attach the charm to the ribbon using a jump ring (see page 26 for how to open and close the jump ring correctly).

3 Glue the bottle and dish together and allow to dry completely.

art supply center

I've thought of all kinds of uses for muffin tins and I buy them by the dozen from my local dollar store to hold beads, buttons, and other jewelry parts. The plastic cups are re-useable over and over, yet can be replaced as needed and best of all they are inexpensive to decorate. Make yours as fun as you would like, or better yet, enjoy this project with your child and let them join in on the fun too.

- 6 plastic cups
- Fabric strips or ribbon
- Self-adhesive magnets
- 6-cup muffin tin
- Acrylic paint
- Paintbrush
- Rickrack trim
- Tacky glue

1 Wrap each cup with a strip of fabric or ribbon and knot in the front.

just a note:

Decorate the cups with decoupaged paper (see page 24) instead of fabric strips if you prefer.

Alternatively you can label the cups with tags, or add chalkboard labels (see page 77).

2 Attach a self-adhesive magnet to the bottom of each cup, to hold it firmly in the muffin tin.

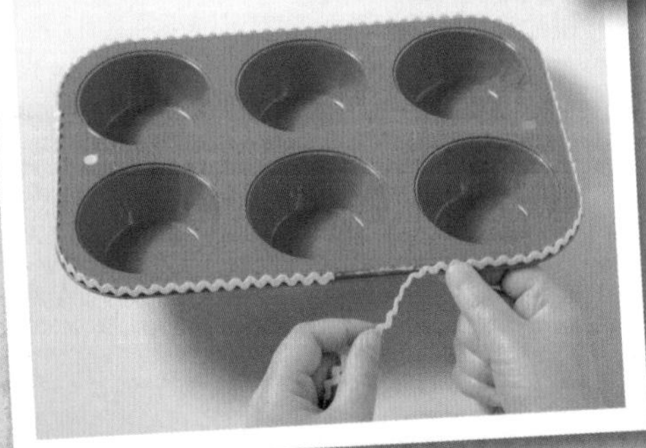

3 Paint the tin in your chosen color and allow to dry. Glue a line of rickrack trim around the edge of the tin. Fill the cups with a selection of your favorite supplies.

mason jar organizer

I especially love using vintage jars because they have so many uses and are artful. This organizer, no doubt, will find a variety of uses throughout your home or office. It's great to keep cotton swabs, cotton balls, and other bathroom toiletries organized, and in your studio to store buttons and smaller beads. If you can't find a suitable Mason jar, simply substitute any jar with a large opening—the clamps come in a variety of sizes and can be found at your DIY home improvement store.

- 2 x 12 in. (5 x 30 cm) piece of wood
- Paint
- Paintbrush
- Chicken wire
- Hammer
- Drill and ¼ in. (6 mm) bit
- 2 galvanized screw collars
- Sanding block
- Brown ink
- Cosmetic sponge
- 2 wood screws
- Screwdriver
- 2 screw eyes
- 2 pint-size (16 fl oz) Mason jars

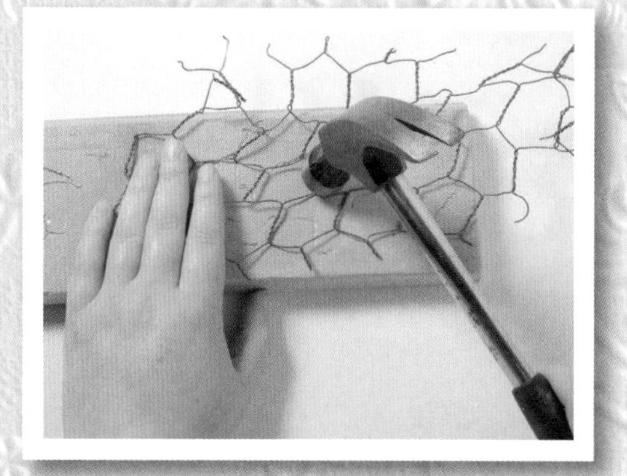

1 Paint the wood in your desired color. Place a piece of chicken wire over the top and hammer lightly to distress and add some interesting pattern to the surface.

2 Drill a hole in the back of each screw collar so that the screw knob will be centered at the front. Drill a pair of matching holes in the wood base. Sand and ink the edges of the wood lightly to distress and highlight the patterns.

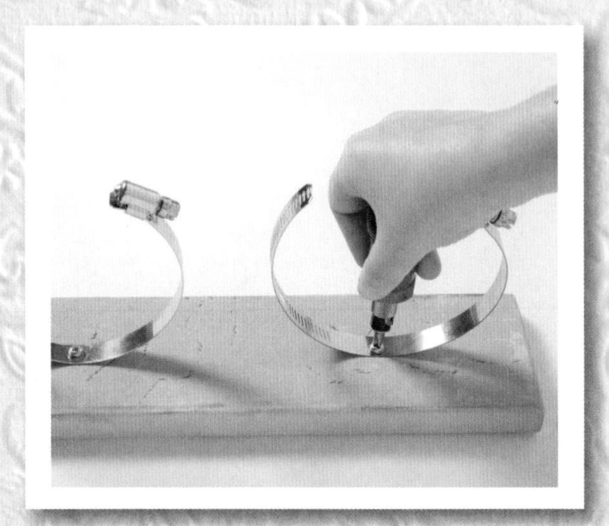

3 Screw the collars to the wood base.

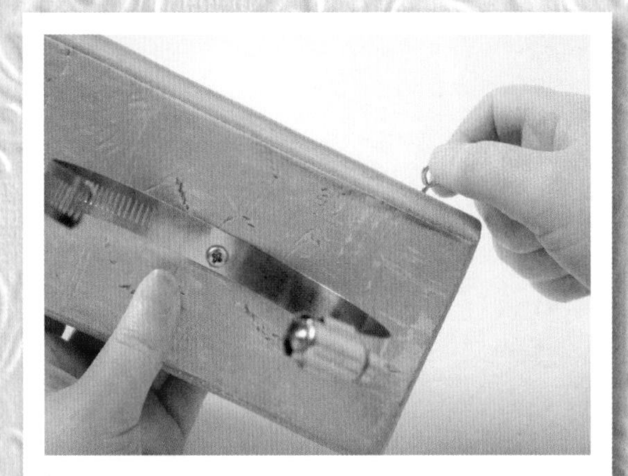

4 Add a screw eye at each end on the top edge to hang the organizer.

just a note:

The screw collars are the type generally found in the heating/cooling area at DIY stores—they are used to clamp ductwork together.

5 Open a screw collar and insert the jar, then tighten the screw knob with a screwdriver. Repeat with the other jar.

basket liners

Fabric-lined baskets in upscale stores are often rather pricey—but you can create your own basket liners to get the same effect for a lot less, and in fabrics to suit your taste or décor. You could also use the techniques in this project to make a fabric liner for the Wire Basket on page 74.

1 Measure the height of the basket and add ½ in. (1.25cm); this will be the depth of the inner strip. Measure around all the sides of the basket and add ½ in. (1.25 cm); this will be the length. Cut a strip of fabric using these measurements. Cut another piece of fabric the same length but only 4 in. (10 cm) deep for the edging. Fold over along one long edge of both pieces by ½ in. (1.25 cm) and stitch a hem.

- Basket
- Tape measure
- Fabric in two colors or patterns
- Scissors
- Sewing machine and thread
- Iron
- Scraps of fabric
- Buttons
- Fabric glue

2 Place the two strips right sides together with raw edges aligned and stitch together along the raw edges, with a ½ in. (1.25 cm) seam allowance. Turn over and press the seam allowance downward.

3 Open up the fabric flat. Place the side ends of the fabric strips right sides together, matching the seam position. Stitch the ends together using a ½ in. (1.25 cm) seam allowance.

4 Turn the fabric piece right side out and set into the basket, turning the edging over the lip of the basket so it runs around the outside top edge.

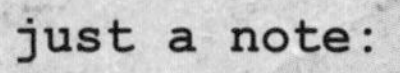

You can turn this into a no-sew project by using fabric glue instead of stitching in steps 1 to 3.

5 Cut graduated circles in different fabrics. Stack and glue together, adding a button on the top. Stick the decoration to the front of the basket with fabric glue, or you could hand stitch in place through the button holes.

flowerpot pens

Dollar store finds get a makeover to turn them into a decorative flowerpot full of blooming pens. You could use this same idea for pencils, or just use popsicle sticks to make an attractive display.

- Ink pens
- Double stick tape
- Ball of twine
- Color spray in green
- 5 x 1½ in. (12.5 x 4 cm) piece of cardboard
- Scissors
- Short length of cord
- Scraps of fabric
- Button
- All-purpose glue
- Package of black beans or beans in your choice of color
- 4 in. (10 cm) clay pot

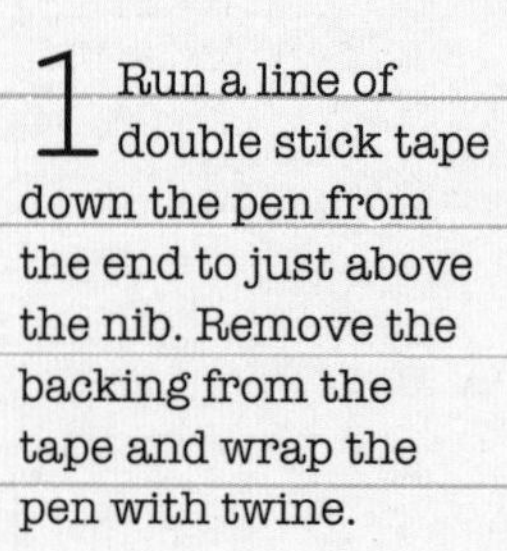

1 Run a line of double stick tape down the pen from the end to just above the nib. Remove the backing from the tape and wrap the pen with twine.

2 Spray the twine with a green color wash—if you want to make your own color wash, see page 29.

just a note:

The flowerpot used here was purchased with a crackle finish, but to achieve a similar effect on an ordinary flowerpot see the crackle glaze instructions on page 22.

Look out for attractive flower pots in your local dollar/pound store or garden center—they are great containers for all kinds of things and you can either use them as they are or add your own additional decoration.

To easily find pens with a different color ink, you could color coordinate the flower center fabric or button to the ink color in each pen.

3 Wrap another length of the twine around the card approximately 20 times. Thread a shorter cord through the wraps, bring the ends together and tie loosely.

4 Remove the cardboard, pull the cord tight and knot. Fluff up the strands into a flower.

5 Flatten the flower and add circles of fabric and a button to the center, using dabs of glue. Cut a leaf shape from green fabric and glue to the back. Glue the flower to the pen base.

6 Add a handful of beans to the bottom of the flowerpot. Tie a length of twine around the flowerpot and decorate with another flower made of fabric circles and a button. Push the pens into the beans.

vintage knob rack

Use this rack to hang necklaces and keep them tidy, place it in the kitchen for small tools, or in the studio for chains and ribbons—the possible uses are endless. The wooden base I used seemed a little boring, so I dressed it up with popsicle sticks and gave it a little texture with chicken wire.

- 11 popsicle sticks
- Wood glue
- Piece of wood approx. 2 in. (5 cm) wide by 12 in. (30 cm) long
- Scissors
- Acrylic paint in base color
- Paintbrush
- Sanding sponge
- Chicken wire
- Hammer
- Antiquing color
- Cosmetic sponge
- Drill and ¼ in. (6 mm) drill bit
- 3 decorative knobs
- 2 screw eyes

1 Glue nine of the popsicle sticks onto the wooden base, using the photograph as a guide. Allow to dry.

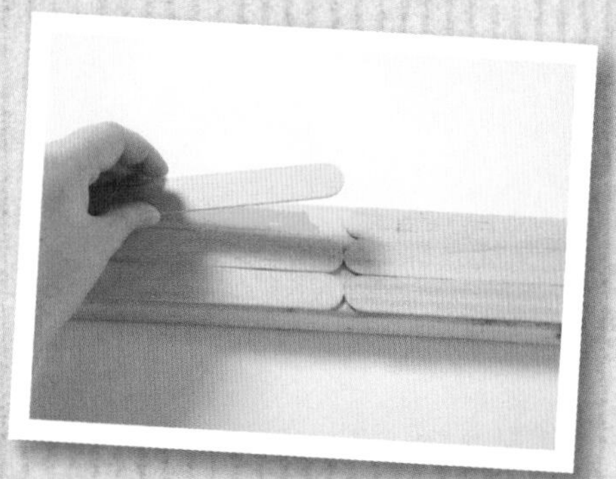

2 Trim the remaining popsicle sticks down with scissors to make two cross bars. Glue these over the joints where the ends of the popsicle sticks meet. Paint the wood and distress as described on page 23.

3 Add antiquing finish as described on page 22. Measure out and drill three holes evenly spaced for the knobs. For extra security, glue in the knobs. Attach screw eyes into the top edge at each end to hang the rack.

just a note:

To make your own antique glaze see the instructions on page 29.

If the screw on the knob is too long, you can cut it down with bolt cutters or use a cutting blade with a Dremel tool. Always wear eye protection when working with these tools.

I purchased these knobs already painted and distressed. Look for knobs that have a fun shape and texture even if they are not the right color—you can always paint and distress them to match your project.

stacked tin caddy

I hate to admit that I often buy products for the packaging. Yes, I like these mints, but what I love is the tins that they come in, which are so useful on so many different levels. Here I've stacked them to make a mini organizing carry-all that can go from place to place.

- 3 mint tins (or similar hinged tins)
- Acrylic paint in rich red and brown
- Paintbrushes
- Crackle medium
- Rub'n'Buff® Gold
- Cosmetic sponge
- Needle tool
- Hammer
- Metal file
- 4 small silver bolts/nuts
- Metal handle
- 4 small brass dome-headed bolts with nuts
- Unmounted rubber stamp with design of your choice
- Black ink pad

1 Paint the base coat on the tin, apply the crackle medium and then apply the top coat (see page 22 for crackle glaze instructions). Apply a little gold Rub'n'Buff® along the edges, using a cosmetic sponge.

2 Stack two tins and open the lid of the top tin. With the needle tool and a hammer, make a hole through the bottom at one side of the top tin and through the top of the second tin underneath. Make another hole on the other side. Repeat with the second tin on top of the third tin. File all the holes smooth and attach all the tins together with the silver nuts and bolts. On the lid of the top tin mark the placement of the handle holes. Punch the holes as above and attach the handle with brass nuts and bolts.

just a note:

If you can't find small enough bolts and nuts in your local hardware store, look for those used in eyeglasses.

You can also substitute brads for the small bolts and nuts, simply adjusting the size of your fixing holes to fit.

3 With all the tins tightly closed, stamp a motif on the front using black stamping ink.

just a note:

We used small tins for our project, but you can use baby jars instead.

Washi tape is printed paper tape that is great for adding quick decoration—find it at your local craft or hobby store.

tiny tins

These are a great way to organize all those small bits and pieces that are so easily misplaced—you can store them in categories so that you can find what you want easily. These can be used as stand-alone storage or used inside the Crafts-on-the-go Trunk on page 98.

- Chalkboard paint
- Tins or baby jars
- Sanding block
- Paintbrush
- Printed paper
- General purpose glue
- Washi tape
- Chalk

1 Make up some chalkboard paint in your desired color as described on page 28. Lightly sand the tops of the tins to create a slightly rough surface so the paint will be able to grip.

2 Paint the chalkboard paint neatly onto the top of each tin and let dry completely. It may take two or more coats.

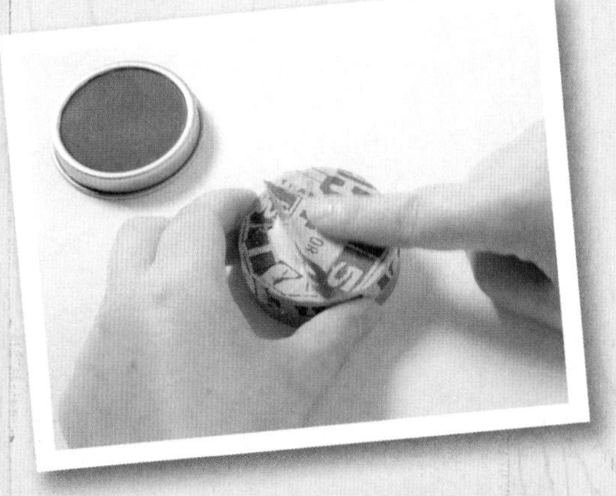

3 Cover the base of the tin with a strip of adhesive washi tape. Label the top of the tin using chalk—you can just rub out the label and re-write when the contents change.

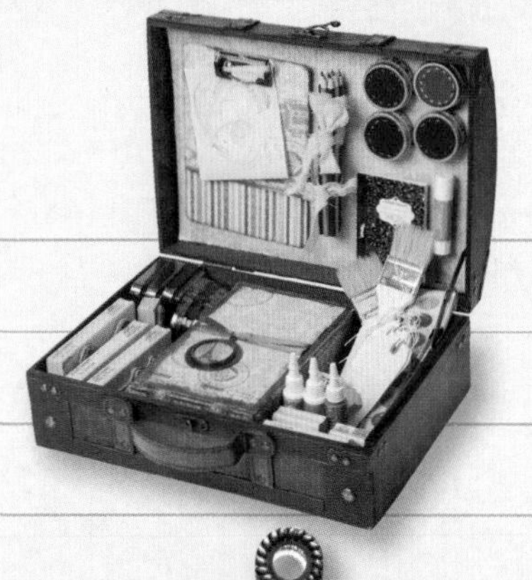

crafts-on-the-go trunk

A true crafter can create anywhere—so make this handy trunk to carry all your crafting gear wherever you need to go. The removable storage tins and boxes make sure that everything has its place but is still easily accessible when you need it.

- Case or trunk
- Tape measure or rule
- Foam board
- Craft knife
- Decorative papers
- Decoupage medium
- Paintbrush
- Miniature clipboard (see page 64)
- Tiny tins (see page 96)
- Notebook
- Hook-and-loop tape
- Magnet strips
- Craft glue
- Cosmetic sponge
- Upcycled cigar box (see page 38)

1 Measure inside the lid of the trunk and cut a piece of foam board to size. Decoupage decorative paper onto one side of the board (see page 24). Decide how the clipboard, tins, and notebook will be positioned on the board. Apply hook-and-loop strips to hold the clipboard, notebook and glue stick in place. Add two magnet strips to hold the tins.

2 Insert the board inside the top of the trunk and secure with craft glue. Run a line of glue around the edges and smooth it with a cosmetic sponge.

3 Attach the clipboard, tins, notebook, and glue stick. Add the cigar box to the trunk and fill the remaining space with your desired contents.

COMPOSITION
WORLD
craft smart
2"
50.8mm

simple solutions

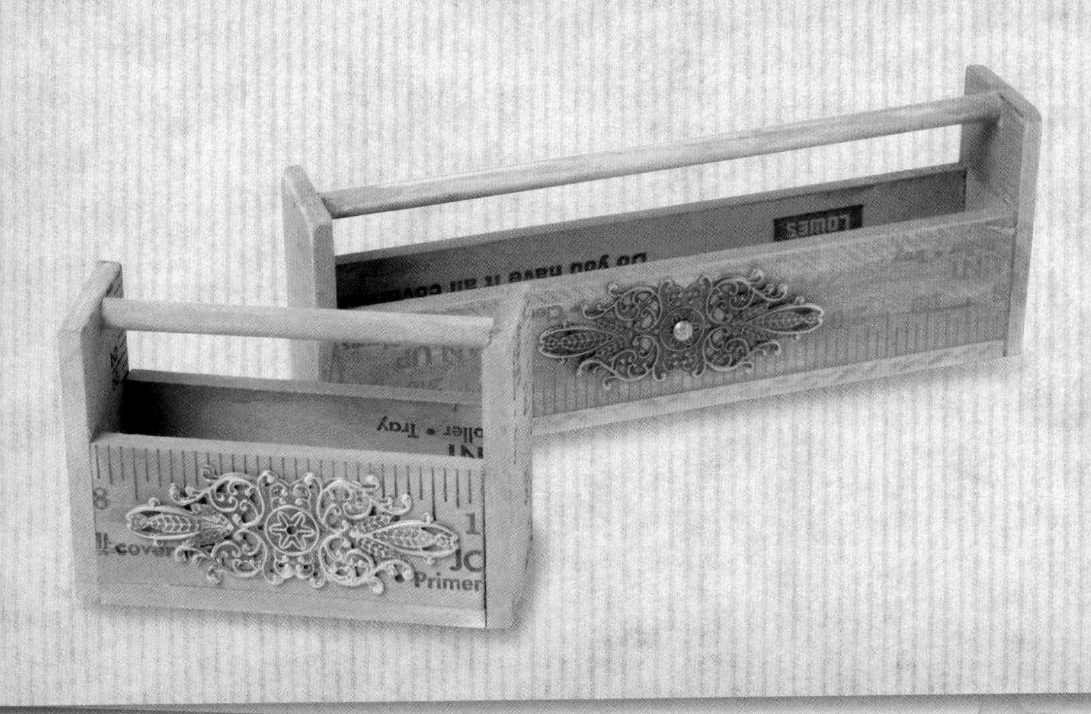

twistie ties

These ties are such an essential part of life to secure almost any type of storage bag, but the plain colors are not very exciting and are sometimes too small to tie bulky items together. Making your own twistie ties means you can make them in all sorts of fun patterns and the length you need.

- Scraps of decorative paper
- Double stick tape
- 26-gauge wire
- Clear packing tape (optional)
- Scissors

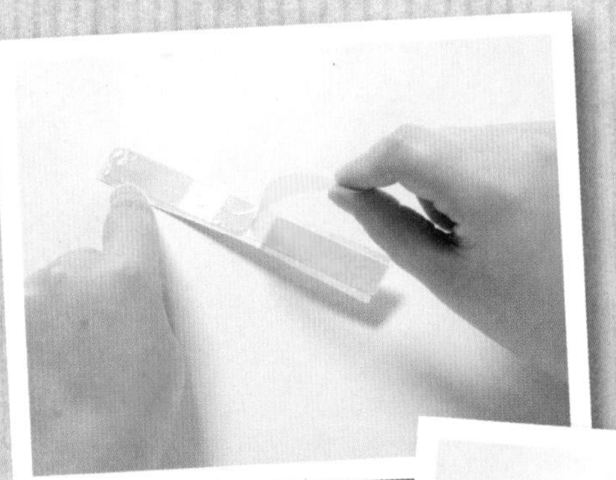

1 Apply a line of double stick tape to the rear of the paper and remove the backing.

2 Apply a length of wire along the center of the tape.

3 Place the remaining paper over the wire and tape, and press down to secure. Cover with packing tape to protect the paper and add extra strength if desired.

just a note:

Twistie ties can be made in a variety of lengths depending on the bulk of the item you are securing.

Photocopy graph paper onto the back of the decorative paper so you can use the lines as a guide to cut out the twisty strips.

As a variation, use patterned Duck Tape® instead of decorative papers. You can also substitute decoupage medium for the packing tape, applying it with a make up sponge to the paper to protect it.

If you want to make several ties at a time, cut decorative paper the size of the double stick sheet, remove one backing from the sheet and stick the paper down. Remove the other backing, lay strips of wire down, followed by another sheet of patterned paper over the top. Cut the twistie ties apart between the wires using scissors.

4 Cut the twistie ties to your desired width and length with scissors.

seaside tumbler

This seaside-theme tumbler is the ideal storage place for makeup brushes or small tools—the sand in the base will keep them upright and securely in position.

- 100 in. (250 cm) twine
- Heavy glass tumbler
- Fine white sand
- Few medium-size seashells
- Glue
- Scissors

1 Fold the twine so you have four strands. Fold all the strands over in half and cross one end over the other on the left side, as shown.

2 Bring the bottom end around and over the top of the other end to make a figure of eight shape, as shown.

3 Thread the strands on the right through the loop on the right.

4 Thread the strands on the left up through the loop on the left. Thread the strands on the right up and over and through the loop on the left.

just a note:

Substitute small pebbles and acorns for a more organic rustic look.

For an organizer for small kitchen tools, use black beans, rice, beans, or split peas in the tumbler.

5 Thread the upper strands on the left up and over and through the loop on the right. Pull all the loops gently to even up the shape and tighten the knot but leave it slightly loose.

6 Add the sand and seashells into the bottom of the tumbler. Wrap the ends of the knotted twine around the tumbler and secure neatly at the back with glue. Trim excess ends.

electrical cord wrap

You'll find so many uses for these tubes that you'll never throw one away again. Here I've upcycled them to tame all my electrical cords. They even work well for my extra long cords that I hide behind my computer—it really tidies up my workspace.

- Cardboard toilet paper tubes
- Scissors
- Double stick tape
- Fabric or decorative paper
- Hook-and-loop tape

1 Cut the tube down the center and open out flat. Apply double stick tape completely over one side and just around the edges on the other. Use the flattened tube as a template to cut out a piece of fabric the same size, and a second piece with an extra ½ in. (1.25 cm) border all around.

2 Remove the backing from the double stick surface of the card and smooth the larger piece of fabric over the top, right side up. Turn the fabric and card over and remove the backing from the double stick strips. Clip across the corners of the fabric and then fold the excess fabric border onto the tape and press in place.

just a note:

I prefer to use fabric that has a little stretch to it instead of paper for the covering, since paper wears over time.

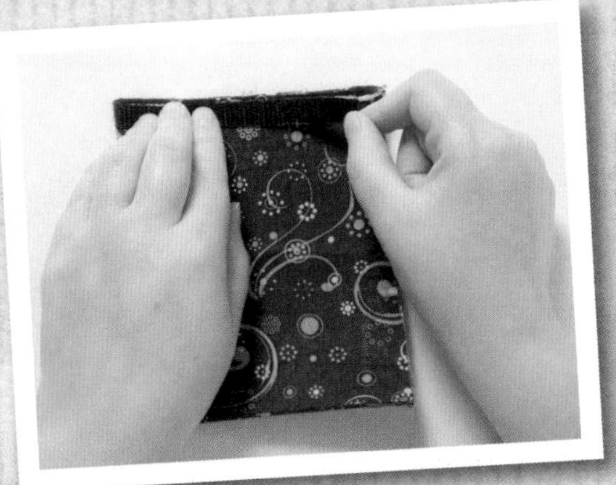

3 Apply double stick tape over the entire surface and add the second piece of fabric on top. Use fabric glue to attach hook-and-loop tape on either side.

Cleopatra's needle

I don't know if Cleopatra used hair ties, but if she did she might have had an organizer for them just like this! And if you don't use hair ties yourself, you can use it to store bracelets instead.

1 Drill a hole at the top of the wooden spool, if your spool does not already have a hole. Add a little wood glue to the bolt of the die and push it into the hole at the top of the spool. Allow to dry.

2 Glue the napkin ring to the bottom of the spool using a little more wood glue. Allow to dry.

- Vintage wooden spool
- Drill and ¼ in. (6 mm) drill bit
- Wood glue
- Die knob with bolt
- Vintage style napkin ring
- Twine
- Vintage style knob with back plate

just a note:

I used a die that was already made into a drawer knob with a bolt fixing, but you can adapt an ordinary die by drilling into one side, inserting a dowel screw and securing it with glue.

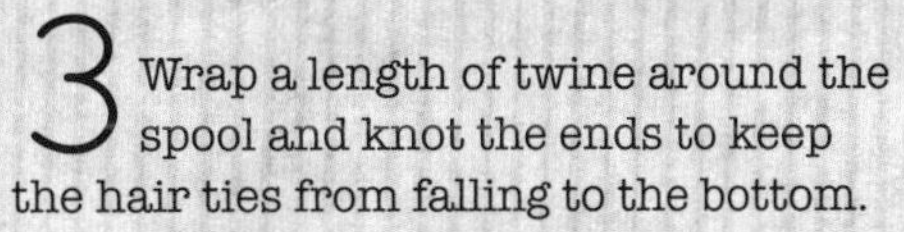

3 Wrap a length of twine around the spool and knot the ends to keep the hair ties from falling to the bottom.

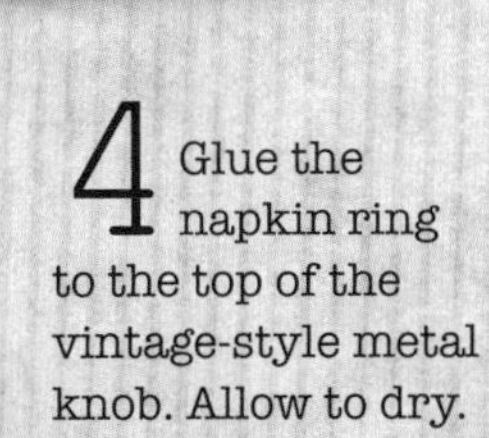

4 Glue the napkin ring to the top of the vintage-style metal knob. Allow to dry.

cable tidy

These up-cycled clips are a great way to keep your power cords from falling off the desk and onto the floor. I've seen these used as is—but I thought they needed to be dressed up a little bit.

- Large binder clips approx. 2 in. (5 cm) wide
- Rule
- Double stick tape
- Decorative paper
- Scissors
- Sanding block
- Decoupage medium
- Cosmetic sponge

1 Measure the front and base of the binder clips and cut two front pieces and one base piece to match from the double stick tape and decorative paper. Apply double stick tape to the front, back and base of the clip.

2 Remove the backing from the tape on the front and back of the clip first and apply the pieces of decorative paper. Rub down firmly along all edges.

just a note:

There are so many ways you can decorate these clips—add some bling with adhesive-backed rhinestones or a little glitter glue.

Label each clip, or color code the clips with acrylic paint according to type of cord or device, so you can easily identify different cords.

If you stand the clips on their base, they double as adorable photo or note holders too!

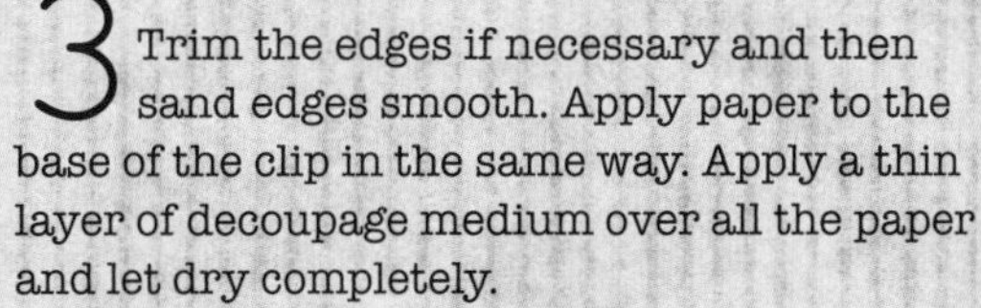

3 Trim the edges if necessary and then sand edges smooth. Apply paper to the base of the clip in the same way. Apply a thin layer of decoupage medium over all the paper and let dry completely.

floral thumbtacks

Decorated thumbtacks and pegs can easily be found in office supply stores, but I find that they can be a bit expensive. Have fun decorating your own—you can make them into anything you like and any color that fits your style and décor. Silicon ice trays work well as molds and come in a variety of shapes and sizes; these can be found at your local kitchen store or party supply store.

- Cool2Cast fiber plaster
- Sealable plastic bag
- Scissors
- Mold
- Thumbtacks/pushpins
- Sandpaper
- Acrylic paint
- Paintbrush
- Glossy decoupage medium or glossy acrylic sealer

1 Mix up the fiber plaster according to the manufacturer's directions. Pour into a sealable plastic bag and massage the mixture well between your fingers to make sure there are no lumps. The mixture should have a pancake batter consistency.

2 Cut off the corner of the plastic bag and squeeze the main body of the bag to fill each opening in the mold. Allow it to sit for a few minutes to thicken.

3 Set the head of a thumbtack into each molded shape. Pour a small additional amount of plaster over the head of the thumbtack to secure, but don't add too much extra—you need to leave a good length of the thumbtack pin free. Allow to dry.

just a note:

You can use two-part resin instead of the fiber plaster. Mix the resin according to the manufacturer's directions, pour into the mold to half fill and allow to set as directed. Set the head of the thumbtack in place and fill the mold with more resin to secure it in position.

You can also use a spray sealer in step 4 instead of the decoupage medium, but make sure you are working in a well-ventilated area.

4 Push out the molded shapes carefully. Sand the edges smooth if necessary and paint the flowers in your desired colors. Seal with glossy decoupage medium or acrylic sealer.

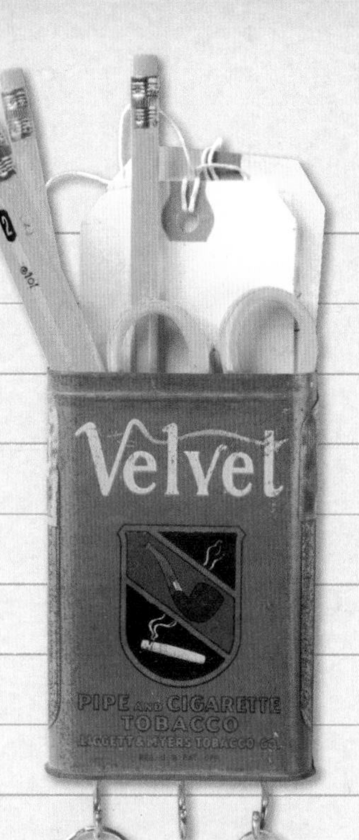

key caddy

Keep keys safe hanging below this useful caddy, which can also be used to hold everyday essentials such as pencils, scissors, and labels.

- Vintage flat can
- Durable needle tool or punch
- Hammer
- 3 screw hooks
- 4 sticky-back magnets

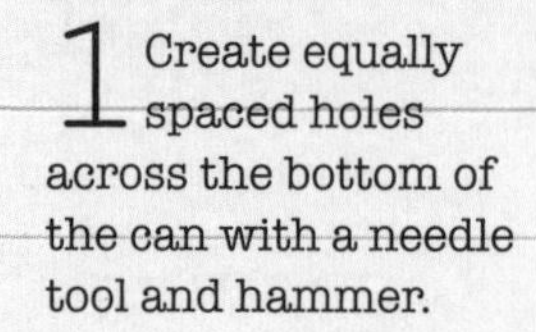

1 Create equally spaced holes across the bottom of the can with a needle tool and hammer.

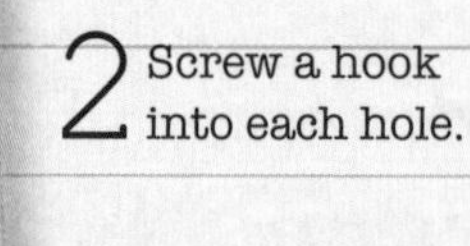

2 Screw a hook into each hole.

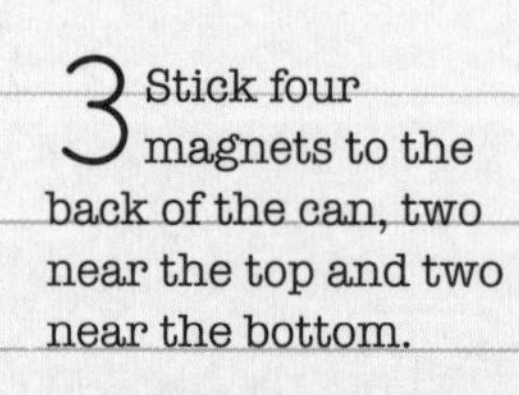

3 Stick four magnets to the back of the can, two near the top and two near the bottom.

just a note:

If your magnets are not the sticky-back type, attach them to the back of the can with all-purpose glue.

cellphone caddy organizer

When your phone is charging, the cord often drapes over the floor—and I never seem to remember where I left mine to charge either. This useful caddy solves both these problems: just pop the phone inside along with the excess cable, and then plug through the back into the wall outlet. The caddy holds the phone neatly against the wall as it charges and is also easy to spot.

- Plastic bottle
- Permanent marker
- Craft knife
- Sanding block
- Tape measure
- Approx 8 x 15 in. (20 x 37.5 cm) of pattern fabric
- Scissors
- Fabric glue
- Cosmetic sponge
- 2½ in. (6.5 cm) square of plain muslin
- Small buttons or embellishments of choice

1 Remove any labels on the bottle. On one side of the bottle draw a large oval or rounded triangle with a permanent marker to mark where the hole to insert the plug through will be. Make sure that this shape is large enough to accommodate your size of plug. Draw a line around the outside of the oval and down and around to the front as a guide to cutting off the top and forming your caddy. Cut away small sections of the bottle at a time, creating smooth curves. Sand the edges if necessary.

2 Measure the circumference and height of cut bottle and cut a piece of pattern fabric to this size plus ½ in. (1 cm) all the way around. Using the cosmetic sponge, cover bottle with fabric glue on the front and back and wrap the fabric around the bottle to cover. Adjust so that the fabric is a snug fit. Apply a little glue to the inside and fold over the edges of the fabric around the low front edge of the bottle, snipping into the edge slightly so it lies flat if necessary.

just a note:

Choose a bottle that is big enough at the base to hold your phone. Use a bottle with a #2 in the recycle triangle—this is usually colored plastic and more durable than clear bottles.

I used an oval bottle around 6½ in. (16.5 cm) high and 3 in. (7.5 cm) across the base—if your bottle is bigger, you may need more fabric.

Make sure your bottle is clean before beginning.

3 Cut away the fabric within the cutout shape for the plug and around the outside of the top curve. Use the bottle as a template to cut a piece of fabric to fit inside the back of the bottle around the plug hole and glue to the inside. Apply glue to the bottom of the bottle and fold the bottom edges of the fabric over the bottom. Draw around the bottom onto the fabric and cut out the shape to glue to bottom.

4 Cut a ½-in. (1-cm) wide strip off the top of the muslin square and fold in half. Create a "bow" by gathering the remaining rectangle of muslin in the center, wrapping it with the narrow folded strip, and securing with a dab of glue. Glue the bow onto the front of the bottle on one side and then glue a small button into the center of the bow.

ribbon sticks

Sometimes it's the simplest of ideas that get overlooked. If you craft or sew regularly, you likely have a large stash of odds and ends of fabric and ribbon stuffed haphazardly into a box. To keep them neat and tidy and to access what you need quickly, try these simple and handy ribbon sticks.

- Tape measure
- Strips of fabric, ribbon, or lace
- Popsicle sticks
- Pencil
- Glass-headed pins
- Glass storage jars

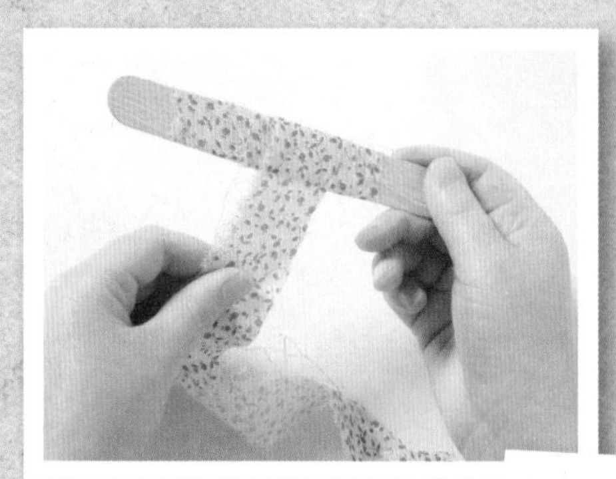

1 Measure the length of the strips before you begin to wrap them around the popsicle stick. Wind each strip around a separate popsicle stick.

2 On the end of the popsicle stick, write down the length it holds. If you write in pencil you can easily change the measurement when you use some of the length.

just a note:

Odd pieces of ribbon are great if you need to add a little extra decoration quickly; you can use them as simple strips of color, tie a bow, or use them to wrap around the project.

You can also use these sticks to store ends of embroidery floss, lace, cords, or twine.

3 Secure the end with a pretty glass-headed pin

4 Sort the sticks by color or pattern and store the different colors in glass jars.

lap dock

Laptops are great when you need a computer you can take around or put out of sight at home. But working at them for long periods of time can cause your hands to hurt since the keyboard is generally resting on a flat surface. This stand is the ideal solution to getting your laptop on an angle to reduce hand fatigue.

- ½ in. (1.25 cm) diameter round PVC pipe or similar
- Tape measure or rule
- Miter/hacksaw
- Sandpaper
- Reel of patterned Duck Tape®
- 6 elbow joints to fit pipe
- Acrylic paint
- Paintbrush
- 2 corks
- Craft knife
- 2 floral thumbtacks (see page 112)
- Glue (optional)

1 Measure and cut the pipe into two pieces 10 in. (25 cm) long, two 2 in. (5 cm) long and one 12 in. (30 cm) long. Sand all the edges smooth. Cover each length of pipe in patterned Duck Tape®, leaving the last ⅜ in. (1 cm) free at each end.

2 Paint the elbow joints in your chosen colors. Push an elbow joint onto each end of the longest pipe, pointing upward. Press a 2 in. (5 cm) pipe into each of these elbow joints followed by another elbow joint on the ends pointing forward. Add the remaining lengths of pipe on each side, and the final pair of elbow joints pointing upward.

just a note:

Adjust the length of the main three tubes to suit the size of your laptop if necessary.

Duck Tape® is available in a wide range of designs and colors, but if you cannot find something you like, paint the tubes instead.

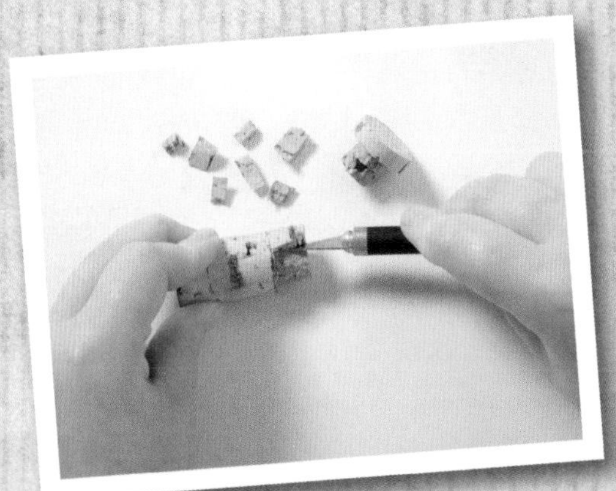

3 Trim down one end of each cork so they will fit into the holes of the last set of elbow joints.

4 Make two Floral Thumbtacks as described on page 112 and glue one into each cork. Place a cork assembly into the hole in each of the final elbow joints and secure with glue if necessary.

mini toolbox

For a long time I have needed a little toolbox for my hand tools and other small items to sit in. I've tried using a bucket or small box, but nothing worked the way I wanted it to. While I was working on a different project for this book, it occurred to me that a yardstick toolbox would be the perfect solution. It's the perfect size to store my note pads and other pieces of station1ry and I've made a separate one to hold my tools perfectly straight.

- Yardstick, approx. 1½ in. (4 cm) wide and ¼ in. (5 mm) thick
- Tape measure or rule
- Miter/hacksaw
- Sanding block
- Wood stain
- Paintbrush
- ½ in. (1.25 cm) brad nails
- Hammer
- Wooden dowel
- Decorative metal plate (brass jewelry finding)
- Small screw
- Screwdriver

1 Cut the following pieces from the yardstick: three 4 in. (10 cm) long pieces (front, back, and bottom) and two 3 in. (7.5 cm) long pieces for the ends. Sand and then stain all the pieces in your chosen color.

2 Align the bottom and side pieces, as shown. Carefully hammer in the brads along the length to secure. When finished with this step you will have a U shape.

3 Attach the ends with brads. Cut a length of dowel to fit between the end pieces and fix in place by hammering a nail brad through each end piece into the dowel.

just a note:

Partly hammer the brad nails into the bottom piece before you start attaching the sides—this will make it easier to hammer them into place.

When hammering the front onto the base, place another length of yardstick underneath in the center for extra support.

4 Attach the decorative plate to the front of the toolbox with a small screw, or you can glue it into place if you prefer.

just a note:

This hanger is designed for pants and it has a double bar with one part that unhooks, so it's ideal to hold your magazine securely on the hanger. If you don't have a hanger like this, it's fine to use one with just a single bottom bar—you could add a Clothes Pin Marker (see page 138) to keep magazines in place on the hanger.

quilted magazine hanger

The ideal storage place for magazines in the bathroom, hang this on the back of the door for easy access to reading materials in the smallest room. Or organize your chains for jewelry making by hanging them from the base.

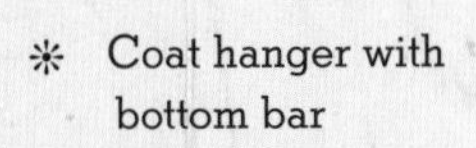

- Coat hanger with bottom bar
- Double stick fabric tape
- Scissors
- Scraps of quilted fabric
- Scrap of lace motif
- Fabric glue
- Raffia

1 Run a line of double stick fabric tape down both arms of the hanger. Wrap each arm with strips of the quilted fabric, snipping it to fit neatly around the hook.

2 Use a little fabric glue to stick a lace motif onto the front of the hanger just below the hook.

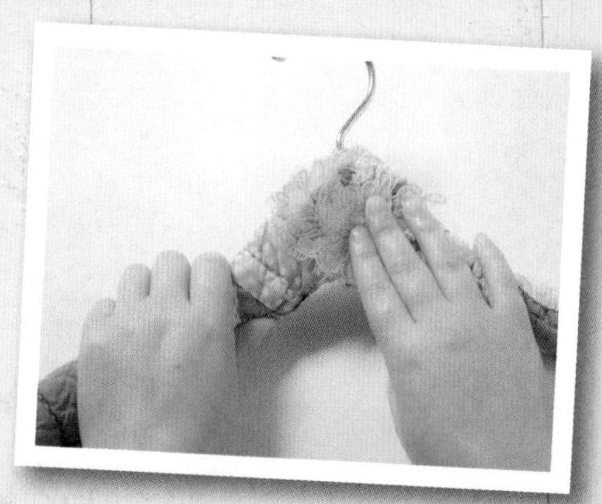

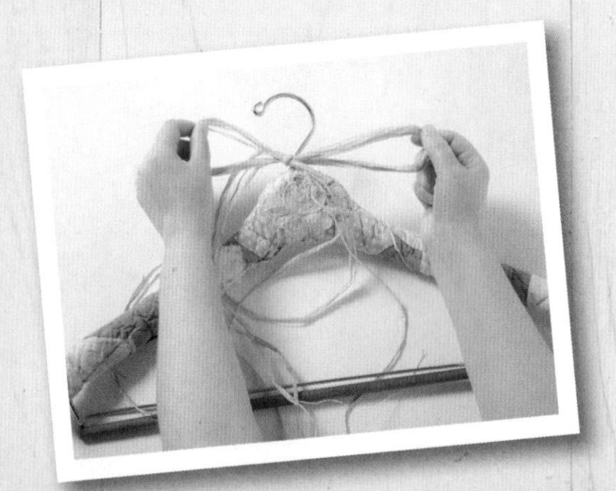

3 Tie a length of the raffia into a big bow on the hook.

magazine bin

Bins like this, for storing magazines or other paperwork, are very useful but can be quite expensive to purchase. Making your own is really simple and you will not only save money but also be able to use your own designs and colors for the decoration.

- 20 x 12 in. (50 x 30 cm) piece of foam board for each bin
- Craft knife
- Dull pencil
- 2 clothes pins
- Glue dot
- Acrylic paint in red and green
- Fast Grab Tacky glue
- Rickrack trim

1 Copy the magazine bin template on page 157 onto the foam board and cut out two pieces. Cut a piece 11 x 2 in. (28 x 5 cm) for the back, another 8½ x 2 in. (21.5 x 5 cm) for the base, and a third 3⅛ x 2 in. (9 x 5 cm) for the front. Cut a small piece of scrap foam board for each stamp and copy the stamp templates on page 157 onto one side. Cut the design into the surface using a dull pencil.

2 Make a handle for each stamp by attaching a clothes pin on the back using a glue dot.

3 Stamp the design onto one side of all the pieces—make sure to stamp opposite sides on the shaped side pieces so you have a pair of sides.

just a note:

If you don't want to make your own stamp you can use purchased stamps—or look for everyday items around the home that would stamp interesting shapes, such as a bottle cap or half a bell pepper.

4 Run a line of glue along the edges of the side pieces and carefully add the back, base, and front to assemble the magazine bin. Leave to dry completely.

5 Run a line of glue all around the front edge and stick on a length of rickrack trim.

E. HAMEL
à Paris
TICKET
Expedition

chapter 5

time management

folding travel journal

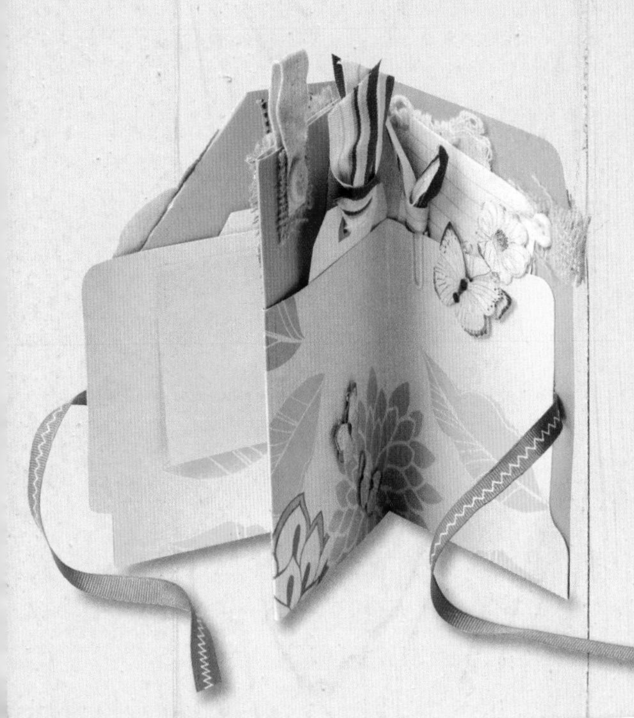

So easy and quick to make, you'll want to make several of these at a time. They are perfect for keeping receipts, managing coupons, or even to keep print souvenirs from a recent vacation.

- Patterned file folder
- Craft knife
- Approx. 24 in. (60 cm) of ¼ in. (6 mm) to ½ in. (1.25 cm) wide ribbon
- Tacky glue or double stick tape
- Pieces of fabric, buttons, stickers, and other embellishments of choice (optional)

1 Open the folder like a book so the inside is facing upward. Fold the bottom edge up toward the top edge by approximately two thirds of the height and crease.

2 Fold the entire piece in half backward.

Journal

just a note:

Decorate your folder with stickers, fabric, buttons, and other embellishments. The front is decorated with the flowers as made in the Basket Liner (see page 85) and with a smaller version of the Artwork Display Frame (see page 70).

I've also added tags and notebooks for notes and some Ribbon Paper Clips (see page 150) to hold items in place.

3 Fold the two sides back toward the front center fold. Mark the position of a slit approx. 2 in. (5 cm) up from the bottom and the width of the ribbon on each of the two side folds.

4 Open the folder completely flat and cut the slits open with the craft knife. Turn over and fold up the bottom again as in step 1.

5 From the outside, thread one end of the ribbon into the right slit, between the layers and out at the right side. Repeat with the other ribbon end on the left. Pull the ribbon ends to tighten and form the book. Glue the layers together along each outer edge, holding the ribbon in place and creating inner pockets.

framed pocket organizer

This is another idea for using air space and getting your items up off your desk or worktop. This pair of pocket frames is combined with a third frame that has a plain fabric panel to use as a pin board. Hang the trio anywhere in the house and you will soon be completely organized. You could create your own fabric patterns by using the stencil technique as detailed on the Retro Memo Boards (see page 144).

- Approx. 40 x 13 in. (100 x 32.5 cm) printed fabric for each pocket organizer frame
- Tape measure
- Pins
- Sewing machine and coordinating thread
- 2 picture frames each 16½ x 13½ in. (42 x 34 cm), or to fit a 14 x 11 in. (35 x 27.5 cm) picture
- 2 pieces of foam board each 14 x 11 in. (35 x 27.5 cm)
- Double stick fabric tape
- Scissors
- 40 in. (100 cm) of 1½ in. (4 cm) wide organza ribbon
- 1 picture frame 18½ x 9½ in. (47 x 24 cm)
- 1 piece of foam board 16 x 7 in. (40 x 17.5 cm)
- 20 x 12 in. (50 x 30 cm) piece of printed fabric

1 Lay out the fabric and measure up one long side, placing a pin at 4½ in. (12 cm), 13 in. (32.5 cm), 20 in. (50 cm), and 35 in. (87.5 cm) up from the bottom. Repeat on the opposite side. Bring the pins at the 4½ in. (12 cm) mark up to meet the pins at the 13 in. (32.5 cm) mark, creating a fold. Press the fold gently by running a fingernail along it.

2 Take this top edge and fold it down over to the front by ½ in. (1.25 cm) to create the double fold. Finger press in place. This makes a wide pocket with a double-folded edge.

just a note:

The taller frame is made in a similar way, but just has a flat piece of fabric over the foam board so it can be used as a pin board.

3 Open up the pocket without disturbing the double fold edge. Stitch along the bottom edge of the double fold to secure it in place. Repeat the same process at the 20 in. (50 cm) pin mark, bringing up to meet the pin at the 35 in. (87.5 cm) mark. Make and stitch the double folded edge in the same way.

4 Fold the pockets up again and then create smaller pockets by stitching a line vertically approx. one third of the way from the right side edge on both top and bottom pocket.

5 Sew right down each side of the panel to hold the pockets in place at the sides. Turn the fabric panel over so it is face downward. Remove the glass and backing panel from the frame.

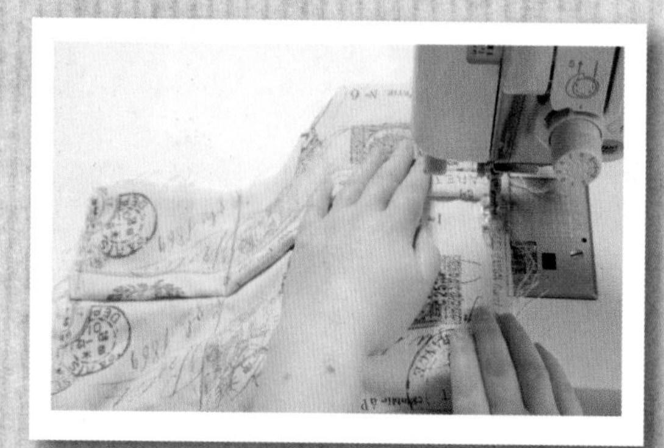

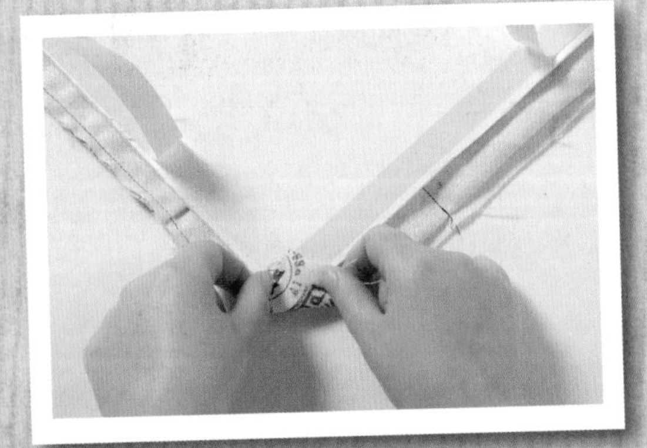

6 Trim a piece of foam board to fit the frame if necessary and apply double stick fabric tape around the outside edges. Place the foam board onto the back of the fabric pocket piece and fold over each corner of the fabric onto the tape.

7 Fold the sides of the fabric panel over onto the tape and press down firmly.

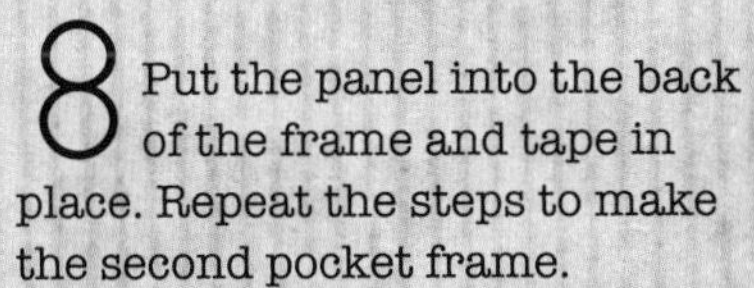

8 Put the panel into the back of the frame and tape in place. Repeat the steps to make the second pocket frame.

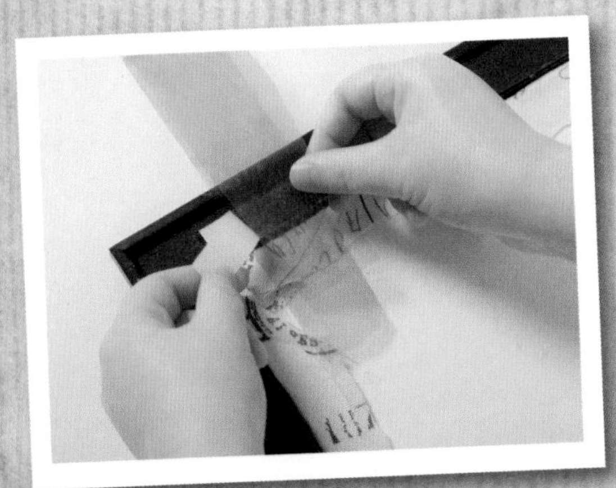

9 Cut two lengths of organza ribbon for hanging loops on each frame and tape in place at the top of each frame on either side. If you prefer, you can cover the backs of the frames with fabric or paper to finish.

clothes pin markers

Who said clothes pins were only for hanging up laundry? This project is way more fun than laundry and so easy too! It's a very simple idea and so quick to make—you'll have loads of fun, decorative clips in no time.

- Clothes pins
- Decorative paper
- Scissors
- Decoupage medium
- Paintbrush
- Sanding block
- Twine
- Thumbtacks

1 Cut pieces of decorative paper slightly wider and longer than a clothes pin. Apply decoupage medium to the back of the paper and one side of the clothes pin.

2 Place the paper onto the clothes pin and smooth out with a brush. Add more decoupage medium on top. Allow to dry. Repeat on the other side of the clothes pin. Decoupage all the remaining clothes pins.

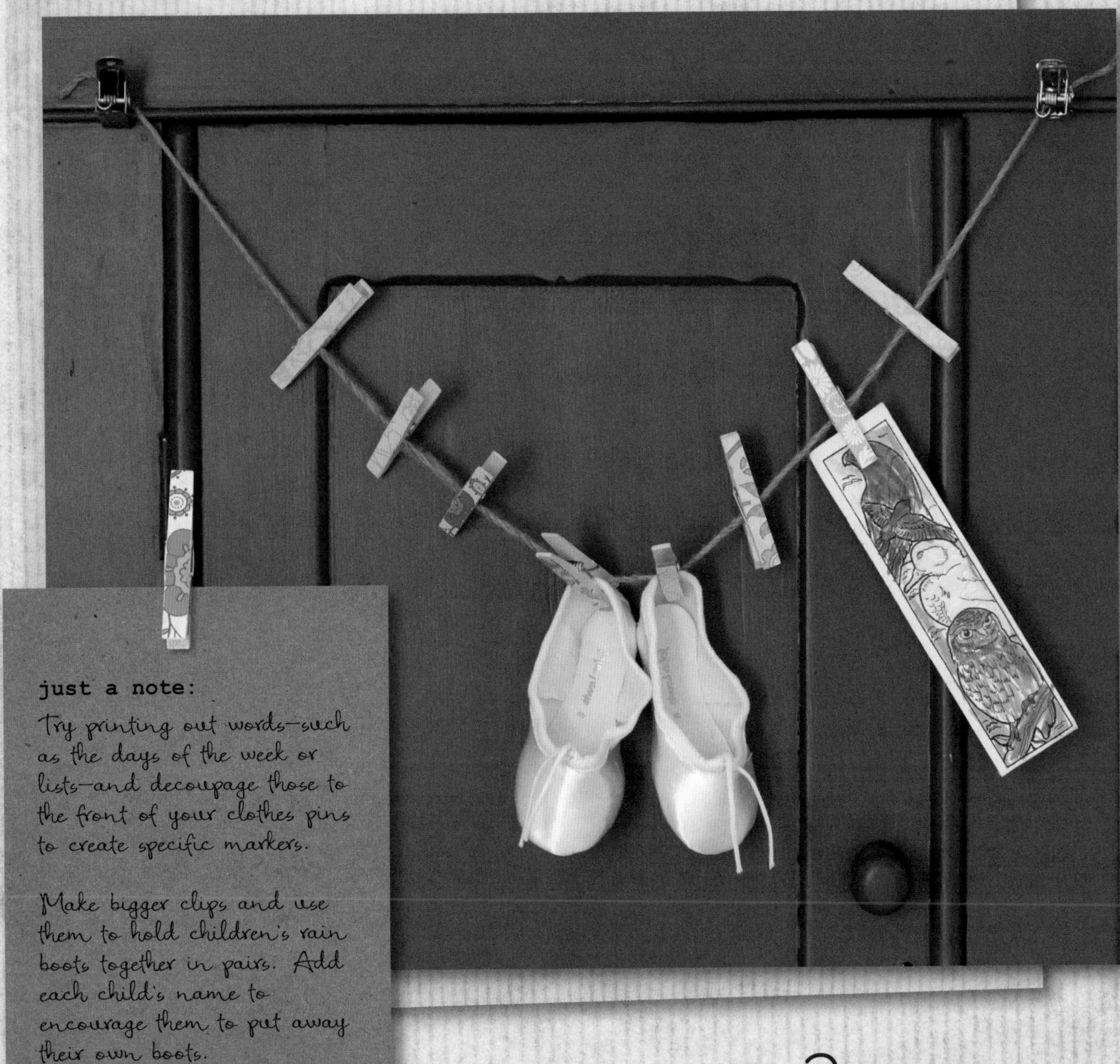

just a note:

Try printing out words—such as the days of the week or lists—and decoupage those to the front of your clothes pins to create specific markers.

Make bigger clips and use them to hold children's rain boots together in pairs. Add each child's name to encourage them to put away their own boots.

3 Sand the edges. Cut the twine to the desired length, knot to create a loop on each end, and use decorative thumbtacks to attach it to a wall or corkboard. You can clip notes to the clothes pins and hang them from the twine.

weekly planner

Glass becomes a dry erase board—this works on all sorts of frames with glass so make a wide variety and use them decoratively in your home, art studio, or office. This board has so many options—you can use it as a message board instead of a planner, replacing days of the week with family names. Send each other positive messages and keep up with daily activities.

- Collage photo frame with at least seven sections
- Tape measure or rule
- Fabric
- Cereal box
- Scissors
- Tacky glue or spray adhesive
- Clear acetate
- Permanent marker
- Alphabet stamps (optional)
- Permanent ink stamp pad (optional)
- Ball of twine
- Scraps of fabric
- Button
- Dry erase markers

1 Remove the backing, glass, and cardboard from the frame. Cut a piece of fabric and a piece of cereal box to fit each opening in the frame. Glue the fabric to the cereal box.

2 Mark the size of the openings on a piece of acetate and write a different day in each box with a permanent marker. Alternatively you can stamp words with alphabet stamps. Cut out the acetate pieces.

3 Insert the glass, acetate, and cereal board/fabric panel into the frame. Finish by replacing the frame back panel. Make a burlap flower (see page 90) and use to embellish the frame.

just a note:

Make sure to stick the fabric onto the un-printed side of the cereal box cardboard. Otherwise the images may show through.

You could use ink jet transparencies for the text by printing your words using a photo-editing program. Set the page dimensions to match the size of the individual frame. Position word text where desired in the frame. Test positioning on plain paper before printing your words on the ink jet transparencies.

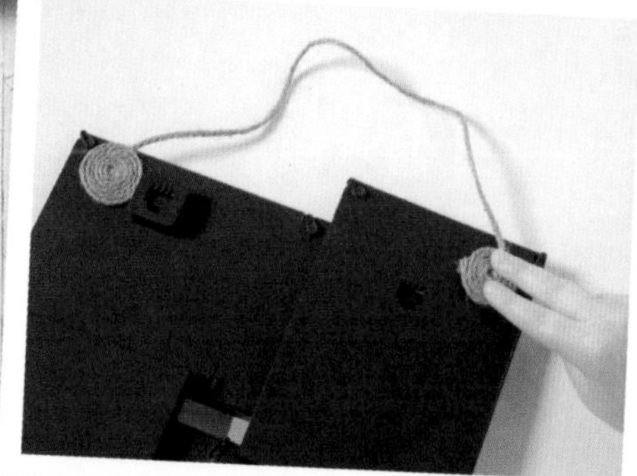

4 If your frame does not have a hanger, you will need to make one. Cut two circles of double stick tape and remove the backing from one side. Curl a length of the twine into a flat spiral on top of each, then remove the backing from the tape and apply the spirals onto the back of the frame with a loop of twine between them.

chalkboard

This whimsical chalkboard can be used anywhere, ready to take notes. Try it in the kitchen or the studio.

- Wooden frame or suitable picture frame
- Acrylic paint
- Paintbrushes
- Chalkboard paint
- Sanding block
- 22-gauge steel wire
- Wire cutters
- Length of dowel
- Pliers
- Drill and 1/16 in. (1.5 mm) drill bit
- Approx. 12 in. (30 cm) length of yardstick
- 2 pieces of chain each approximately 1½ in. (4 cm) long
- Silicone-based glass glue
- Wood glue

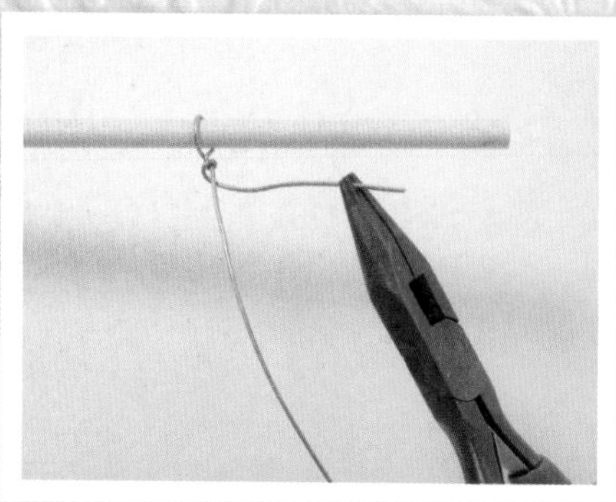

1 Paint the frame in your desired color. Remove the glass from the frame and paint with chalkboard paint (see page 28). Distress the edges of the frame if desired (see page 23). Cut four pieces of wire each 2 in. (5 cm) long. Bend a piece of wire around the dowel to form a loop and twist one wire end around the base of the loop to create a screw eye. Repeat to make three more screw eyes.

2 Drill a hole on either side of the frame near the bottom and on each end of the top of the length of yardstick rule. Trim the end of each screw eye to ¼ in. (6 mm), insert one into each drilled hole and bend the end around with pliers so that it digs in to secure.

3 Attach one end of a piece of chain to each screw eye in the frame by opening up the links with pliers. Attach the opposite end of the chain to the screw eyes in the rule.

just a note:

I purchased this delightful laser-cut frame from a craft store and recycled a piece of glass taken from a broken frame to go inside.

Any size frame will work for this and small frames make adorable little chalkboards on which to place small notes.

4 Using a silicone-based glue, attach the glass chalkboard to the back side of the frame. If you are using a picture frame, you can insert the chalkboard glass as normal.

5 Apply a thin line of wood glue to the back edge of the rule and attach it to the bottom of the chalkboard frame. Wipe away excess glue with a rag and allow to dry.

retro memo boards

Why should memo boards be rectangular or square? These circular memo boards have a stenciled retro design that will add a touch of style to any room.

- Small, medium, and large foam circle shapes
- Piece of burlap
- Scissors
- Hot glue gun
- Stencil of choice
- Black acrylic paint
- Cosmetic sponge
- Fabric glue

1 Use each foam shape to cut a circle from the burlap that is 2 in. (5 cm) bigger all around. Start folding the excess fabric up and over to the back of the foam shape all round and secure with the hot glue gun.

2 Stencil your chosen design onto the front of the fabric using black acrylic paint and a cosmetic sponge. Allow to dry.

3 Cut another circle of burlap the same size as the foam shape and glue onto the back with fabric glue. Repeat the steps to make the other two boards.

just a note:

Create thumbtacks to match by following the instructions on page 112.

If you don't want to create your own fabric designs, simply select coordinating fabrics to cover the circles.

magnetic memos

Display your notes and pictures with style on this magnetic memo board, which is really quick and easy to make and can be displayed pretty much anywhere. See page 148 for how to make magnetic pebbles to use with this board.

- ✻ Baking tray
- ✻ Sanding block
- ✻ Paintbrushes
- ✻ Acrylic paint
- ✻ Decoupage medium
- ✻ Decorative paper
- ✻ Cosmetic sponge
- ✻ General-purpose glue
- ✻ Rickrack trim

1 Lightly sand the baking tray and then paint the reverse side with acrylic paint in a color of your choice.

2 Paint the reverse side of the baking tray with a layer of decoupage medium.

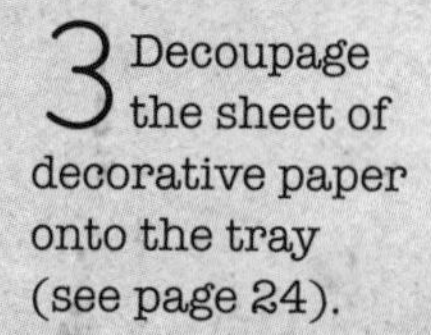

3 Decoupage the sheet of decorative paper onto the tray (see page 24).

4 Run a line of glue around the edges of the decorative paper and attach the rickrack trim in place.

just a note:

If the design on the decorative paper is not directional, you can hang the board either upright or horizontally.

just a note:

Your memo board can be displayed on an easel, but if you would rather hang it up glue a length of ribbon on the back to make a loop. Alternatively, drill two holes into the top edge about 4 in. (10 cm) apart, thread in a strip of muslin, and tie in a knot to secure.

magnetic pebbles

Use these on your fridge or anywhere you can use a magnet—or make your own magnetic memo board by following the instructions on page 146.

1 Remove the covering from one side of a strip of double stick sheet and apply the glass pebbles.

- Double stick sheet
- Glass pebbles
- Scissors
- Decorative paper
- Sanding block
- Magnets
- Tacky glue

2 Cut out each pebble with scissors—the cutting doesn't have to be perfectly neat to the edge at this stage.

3 Remove the backing sheet from the double stick tape on the pebble and apply the pebble over a suitable motif on the paper. Cut out around the pebble with scissors again.

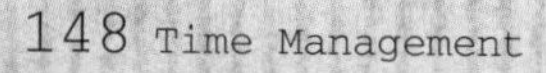

just a note:

Instead of using the designs on decorative paper, place the pebbles over words or names. You can also use miniature photographs for a more personal touch. Just edit the photos down to size in your photo-editing program.

4 Trim the paper flush to the edge of the pebble. To make a nice oval or round circle, use a sanding block to gently polish the edges. Attach a magnet square to the back of each pebble with tacky glue.

ribbon paper clips

The paper clip gets full marks for sheer practicality—but fewer marks for attractiveness. Jazz your paper clips up quickly and easily with these practical ribbon ties. They make great bookmarks too and you can even personalize them!

- ½ in. (1.25 cm) wide strip of decorative fabric
- ¼ in. (5 mm) wide iron-on fusible webbing tape
- Iron
- ¼ in. (5 mm) strip of muslin
- Scissors
- Colored paper clips
- Fine marker pen (optional)

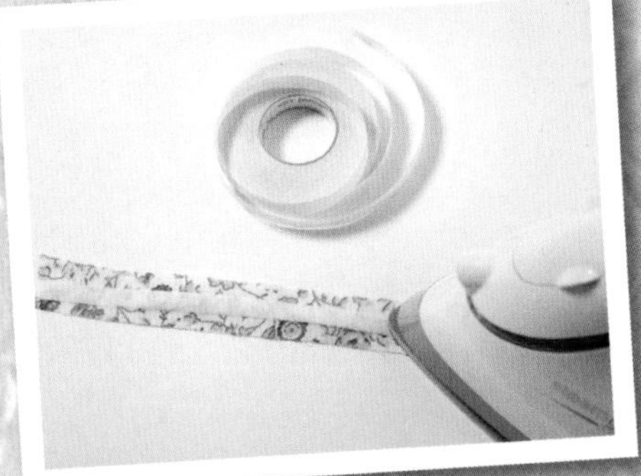

1 Iron the fusible webbing tape onto the right side of the strip of decorative fabric, following the manufacturer's instructions.

2 Remove the paper backing from the tape. Add the strip of muslin on top and iron this in place in the same way.

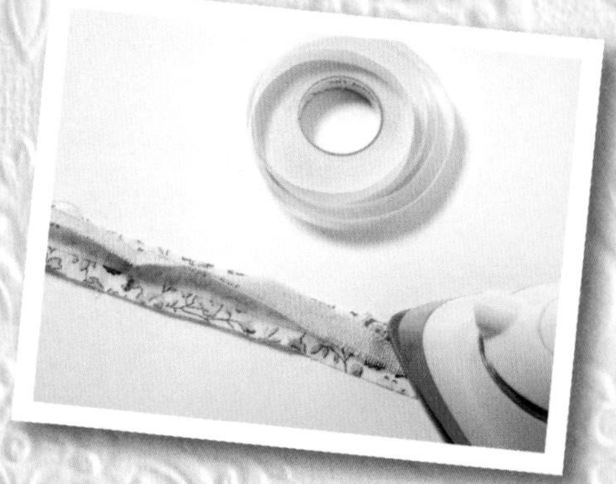

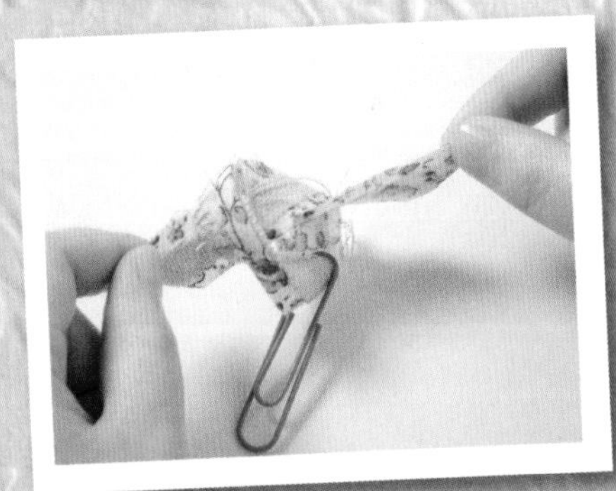

3 Cut the strip into lengths of about 4 in. (10 cm) long. Knot each strip onto the top of a paperclip.

just a note:

Add notes to the muslin strip using a marker pen to highlight what the clip is holding.

flip calendar

What a cute and simple way to organize your day and your thoughts all in one place! For a reusable day calendar simply write the name of the month, days, and dates on each individual notebook and decorate. Or use it as a note calendar with day and date on separate notebooks and use the third to record events and appointments.

- Paint
- Paintbrush
- 3 mini spiral top-bound notebooks
- Dowel rod
- Small saw
- Decorative stickers, fabric, or paper
- Marker
- Strip of fabric/ribbon
- Scissors

1 Randomly paint some of the pages in each notebook and allow them to dry. Cut the dowel so it is slightly longer at each end than the three notebooks laid side by side. Paint the dowel with your desired color.

2 Embellish some of the notebook pages with stickers, fabric pieces, or decorative paper.

3 Thread the dowel through the spirals on the notebooks. To use as a calendar, write Month, Day, and Date at the top of the relevant page.

4 Tie a length of torn fabric to the ends of the dowel. Hang from a nail or tack on the wall or bulletin board.

resources

Linda Peterson

Website:
lindapetersondesigns.com

Facebook:
Linda Molden Peterson

Youtube:
youtube.com/
lindapetersondesigns

Pinterest:
lindapetersondesigns

Email:
lindapetersondesigns
@yahoo.com

US SUPPLIERS

MATERIALS

Suppliers of the special materials used in the projects.

Spellbinders®

Artisan X-plorer™ machine, Spellbinders® die cutting templates, Susan Lenart Kazmer for Spellbinders® Media Mixage™, Bezel Blanks

www.spellbinderscreativearts.com

IlovetoCreate

Collage Pauge decoupage medium, Aleene's® original Tacky Glue®, Fast Grab Tacky Glue®, No-Sew, Fabric Fusion tape, Glass and Bead glue, Tacky Dots®, Wood Glue®, magnet dots, hook-and-loop tape

www.ilovetocreate.com

Duck® brand

Patterned duck tape in rolls and in flat sheets, hundreds of colors and patterns to choose from.

www.duckbrand.com

Cool2Cast™

Fiberplaster casting compound

www.cool2cast.com

Plaid Enterprises

Apple Barrel® acrylic paints, crackle medium

www.plaidonline.com

Black and Decker

Battery-operated cordless drill

www.blackanddecker.com

Beadalon, Inc

Basic jewelry findings, jewelry-making tools, small hand tools

www.beadalon.com

Ranger Inc

Inks, color wash

www.rangerink.com

Jack Richeson

Gesso board, chipboard

www.richesonsart.com

Amaco., Inc.

Rub-N-Buff®

www.amaco.com

GENERAL CRAFT

Craft materials and tools.

Hobby Lobby Stores

www.hobbylobby.com

Michaels Stores

1-800-642-4235

www.michaels.com

A.C. Moore

1-888-226-6673

www.acmoore.com

JoAnn Crafts

1-888-739-4120

www.joann.com

HARDWARE SUPPLIERS

For specialty nuts, bolts, chains, and pipe.

Ace Hardware

www.acehardware.com

Lowe's Inc.

www.lowes.com

Home Depot

www.homedepot.com

UK SUPPLIERS

MATERIALS

Suppliers of special materials used in the projects.

Craft Obsessions

Spellbinders® die cutting templates, Media Mixage™, Bezel Blanks, adhesives, glue dots, decoupage paper, washi tape, inks, chipboard, Mod Podge

www.craftobsessions.co.uk

Amazon

Patterned Duck tape, Rub-N-Buff®, Wundaweb (no-sew tape), magnetic dots, hook-and-loop tape and dots, decoupage materials

www.amazon.co.uk

Craft Mill

Casting materials, molds, air dry clay, polystyrene shapes, decoupage supplies

www.craftmill.co.uk

The Bead Shop

Basic jewelry findings, beads, charms, jewellery-making tools

www.the-beadshop.co.uk

London Graphic Centre

Art supplies, brushes

Three stores in central London

www.londongraphics.co.uk

Cass Art London

Art supplies, brushes

Five stores in central London

www.cassart.co.uk

GENERAL CRAFT

Craft materials and tools.

Hobbycraft

Stores nationwide

www.hobbycraft.co.uk

Crafty Devils

Online store

www.craftydevilspapercraft.co.uk

John Lewis

Stores nationwide

www.johnlewis.com

The Craft Barn

Online store

www.thecraftbarn.co.uk

The Range

Stores across England and online store

www.therange.co.uk

HARDWARE SUPPLIERS

Speciality nuts, bolts, chains and pipes.

B&Q

Stores nationwide

www.diy.com

Homebase

Stores nationwide

www.homebase.co.uk

The publishers would like to thank the following stores and suppliers for loaning props for photography:

Hexagone

French concept store

www.hexagone-uk.com

Folklore

www.shopfolklore.com

Janome

Sewing machines

www.janome.co.uk

templates

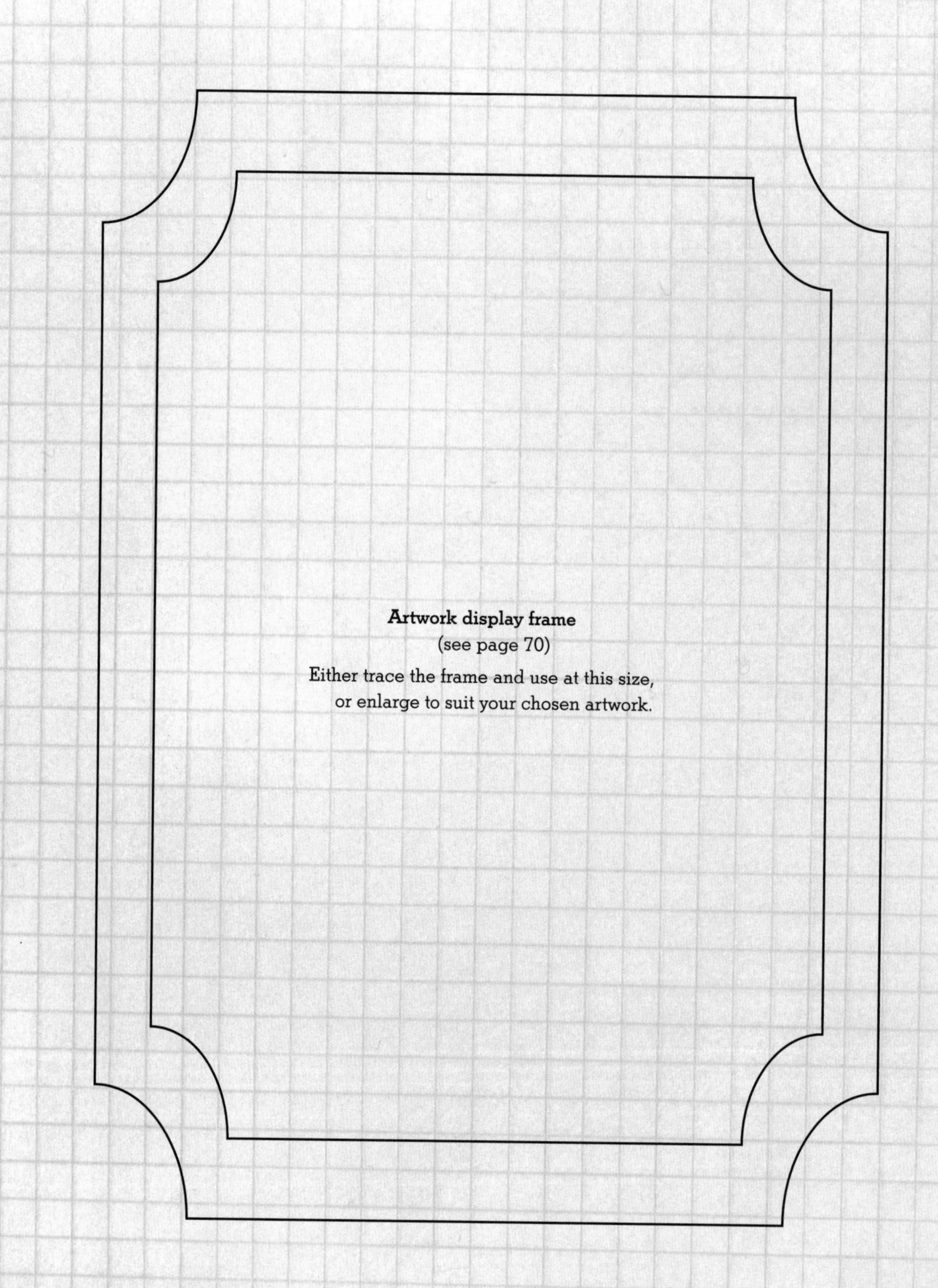
Artwork display frame
(see page 70)
Either trace the frame and use at this size,
or enlarge to suit your chosen artwork.

Magazine bin

(see page 126)

Stamps—trace the motifs and use at this size.

Magazine bin

(see page 126)

Side panel—enlarge by 200% and cut two pieces.

index

acknowledgments

My love and thanks go especially to my husband, Dana; during the creating of this book I worked long hours in the studio and we didn't get much time together. My love and appreciation goes out to him for his self-sacrifice, and for his support and being my best cheerleader.

I could not possibly have accomplished this book without the help of my fantastic team! While my name may be on the front cover, I am only one of many people involved in bringing this book to you.

A special thanks goes out to my publisher, Cindy Richards and her talented team of Sally Powell, Gillian Haslam, and Penny Craig; my personal editor Marie Clayton who always spends endless hours making sure all my "t"s are crossed and my "i"s are dotted; my photographer Geoff Dann, who is brilliant at making my work shine; Marc Harvey who is a tremendous photography assistant; Luis Peral-Aranda for his talent in styling the photos; Gavin Kingcome, the style photographer; and Mark Latter, the book designer. And certainly, not least by any means, special thanks to my daughter Mariah Welsh, who was my very patient hand model.

Each one of these people provides a vast amount of knowledge and talent, I am grateful to have each of you and appreciate the hard work and dedication you have to your craft and this book. You are the best.